CONTROVERSIES IN ENVIRONMENTAL SOCIOLOGY

This comprehensive textbook deals with the key issues and controversies in environmental sociology today. Each chapter deals with discrete issues in a manner that captures the main debates, the central figures, and the social nature of environment-related trends. The text reflects international developments in the area, as well as drawing upon specific case examples and materials. It includes contributions from leading experts in the field, and is compiled by one of Australia's best-known sociologists, Professor Rob White. Written in accessible language, with further reading lists for students at the end of each chapter, *Controversies in Environmental Sociology* provides a timely introduction to the subject.

Rob White is Professor of Sociology at the University of Tasmania.

CONTROVERSIES IN ENVIRONMENTAL SOCIOLOGY

Edited by Rob White

CAMBRIDGE
UNIVERSITY PRESS

CAMBRIDGE
UNIVERSITY PRESS

477 Williamstown Road, Port Melbourne, VIC 3207, Australia

Cambridge University Press is part of the University of Cambridge.

It furthers the University's mission by disseminating knowledge in the pursuit of education, learning and research at the highest international levels of excellence.

www.cambridge.org
Information on this title: www.cambridge.org/9780521601023

First published by Cambridge University Press 2004

A catalogue record for this publication is available from the British Library

National Library of Australia Cataloguing in Publication data

White, R. D. (Robert Douglas), 1956–.
Controversies in environmental sociology.
Includes index.
For tertiary students.
ISBN 0 521 60102 9.
1. Social ecology – Textbooks.   I. Title.
304.2

ISBN 978-0-521-60102-3 Paperback

CONTENTS

TABLES

FIGURES

AUTHOR NOTES

Sharon Beder is a Professor in the School of Social Sciences, Media and Communication at the University of Wollongong.

Lynda Cheshire is a postdoctoral research fellow in the School of Social Science at the University of Queensland.

Lindie Clark is Lecturer in Health Studies at Macquarie University.

Peter Curson is Professorial Fellow in Medical Geography and Head of the Health Studies Program at Macquarie University.

Julie Davidson is currently a Postdoctoral Research Fellow and principal researcher in the Sustainable Communities Research Group in the School of Geography and Environmental Studies, University of Tasmania.

Aidan Davison is an Australia Research Council Fellow in the Sustainable Communities Research Group at the University of Tasmania.

Douglas Ezzy is a Senior Lecturer in Sociology at the University of Tasmania.

Vaughan Higgins is a Lecturer in sociology at the Gippsland Campus of Monash University, Australia.

Natalie Jackson is a Senior Lecturer in social demography at the University of Tasmania.

Roberta Julian is Associate Professor and Director of the Tasmanian Institute of Law Enforcement Studies at the University of Tasmania.

Geoffrey Lawrence is Professor of Sociology, and Head of the School of Social Science, at the University of Queensland.

Stewart Lockie is Director of the Centre for Social Science Research and Senior Lecturer in Environmental and Rural Sociology at Central Queensland University.

Lyle Munro teaches Sociology in the School of Humanities, Communications and Social Sciences at the Gippsland campus of Monash University in Victoria.

Kristin Natalier is an Associate Lecturer in the School of Sociology and Social Work, University of Tasmania.

Val Plumwood is an Australia Research Council Fellow at the Australian National University.

Carol Ackroyd Richards is a PhD candidate in the School of Social Science at the University of Queensland.

Elaine Stratford is a Senior Lecturer in Geography and Environmental Studies at the University of Tasmania.

Bruce Tranter is Senior Lecturer in the School of Political Science and International Studies at the University of Queensland.

Frank Vanclay is a Professorial Research Fellow in rural sociology at the Tasmanian Institute of Agricultural Research at the University of Tasmania in Hobart.

Rob White is Professor of Sociology at the University of Tasmania.

ABBREVIATIONS

AGPS	Australian Government Publishing Service
ANU	Australian National University
COAG	Council of Australian Governments
CQU	Central Queensland University
EIA	Environmental Impact Assessment
EMO	environmental movement organisation
EMS	environmental management system(s)
FAIR	Fairness and Accuracy in Reporting
GEF	Global Environment Facility
GMO	genetically modified organism
IAIA	International Association for Impact Assessment
IPCC	Intergovernmental Panel on Climate Change
ISO	International Standards Organisation
LULUs	locally unwanted land uses
NEPA	*US National Environment Policy Act 1969*
NEPC	National Environment Protection Council
NH&MRC	National Health & Medical Research Council
NIABY	Not In Anyone's Backyard
NIMBY	Not In My Backyard
NNIS	National Nosocomial Infections Surveillance System
NOHSC	National Occupational Health and Safety Commission
NOPE	Not On Planet Earth
NSM	new social movements
PBS	Public Broadcasting Service
PETA	People for the Ethical Treatment of Animals
SEA	strategic environmental assessment
SIA	Social Impact Assessment
SIDS	small island developing states
SMO	social movement organisation
STS	science, technology and society
TFR	Total Fertility Rate
TWS	The Wilderness Society
UNESCO	United Nations Educational, Scientific and Cultural Organization
UNSW	University of New South Wales
USAID	United States Agency for International Development
WSSD	World Summit on Sustainable Development

INTRODUCTION
Sociology, Society and the Environment

Rob White

Environmental issues, problems and struggles are central to human life in the 21st century. The relationship between 'society' and the 'environment' has generated much in the way of both action and analysis over the last thirty years. As local and global environments rapidly change, and as humans modify their behaviour in relation to how and where they live, the importance of studying the interface between society and environment has likewise steadily grown.

Different writers have different conceptions as to what constitutes the most appropriate way to analyse 'environment and society', and indeed what to include as part of such discussions. For some, the important thing is to consider particular environmental issues such as soil degradation, declining biodiversity, solid waste problems, chemical pollution, global climate change, use of fossil fuels – the list goes on. For others, the approach may be more conceptual, in the sense of locating debates over and about the environment within the context of social and political theory, such as analysis of different ways in which 'nature' is defined and perceived, theorising the relationship between human beings and 'nature' and human beings and non-human animals, examining the ways in which industrialisation and globalisation impinge on environments, and exploring the agency of human beings in relation to their environments and as part of social movements about the environment. The complexity and overlap of issues and approaches surrounding the environment means that there will necessarily be myriad different ways in which to study the environment–human nexus.

The aim of this book is to provide an introduction to environmental sociology, and to do so by providing an overview of key controversies within the field. The book is meant to whet the appetite for sociological analysis of environmental issues, to raise relevant questions, rather than to provide definitive answers or one-size-fits-all theoretical models. As such, the intention is to stimulate further thinking and research in this area, and to indicate future lines of sociological inquiry. This is evident in the wide range of issues and approaches discussed in the book. From demographic changes to unhealthy living environments, perceptions of technology to assessment of risk, the chapters present insights into the nature of many different types of environmental issues. They do so by comparing and contrasting competing and often opposing perspectives, thus illustrating the tensions and conflicts in how issues are defined, perceived and responded to. Collectively they demonstrate the varieties, and importance, of environmental sociology.

ABOUT ENVIRONMENTAL SOCIOLOGY

Sociology is about people, institutions and behaviours. It is about social interactions and social structures. Ideally, sociology consists in thinking about the nature of society, and comparing any particular society with what went before and what it is likely to become. The concern is with both 'what is', and 'what ought to be'. The task of the sociologist, in this perspective, is to stand back from commonsense views of the world to investigate where we are and where we are going. It is about gaining a sense of historical and global perspective. It is about understanding the structure and processes of a society as a whole, including global societies.

Sociology is about issues relating to social inclusion – the 'insiders' and participating members of any society. What are the ways in which people behave, feel, think, and act, and why do they do so in the ways they do? What binds us together as a social whole? Sociology is also about issues relating to social exclusion – the 'outsiders' who are excluded from the mainstream of social life and social opportunities. How do we explain why it is that some people are privileged in their lifestyles and choices, while others are disadvantaged? What separates us from each other? Sociology is thus about the boundaries of the conventional and the unconventional. It is about the dynamic ways in which people are brought into and left out of the social order.

Sociology is about understanding and dealing with social problems. It is concerned with defining whether or not an issue is indeed a social problem, rather than simply a personal trouble or a natural phenomenon, and why this is the case. It is also concerned with devising social policies and/or developing practical applied strategies that can be used to address social problems. In other words, sociology is about acting on judgements made about the world around us. In my view, sociology is about putting things into context, about challenging the status quo, and about making the world a better place. It is essentially about three important tasks: *see, judge, act.*

Environmental sociology is about translating these tasks into analysis and action around environmental issues. To illustrate this, we can consider the matter of drinking water. Sociologically, investigation of drinking water could proceed by looking at how water is managed and distributed, historically and in different cultural contexts. It could examine differences and similarities between societies in which drinking water is freely provided, and those in which it is sold for profit. It could compare the place of water in societies in which it is scarce, with those in which it is abundant, from the point of view of control, access and symbolic importance. Social differences in the use of water may be apparent within a society. So too, water may represent affluence for specific classes and castes, or for particular societies compared to others. Water, therefore, is integrally linked to certain kinds of social structures, social interactions and social processes of inclusion and exclusion.

A distinction can also be made between a 'problem' (unsafe drinking water) and a 'sociological problem' (why or why not unsafe drinking water is considered a social problem). In some towns and cities, for example, poor-quality drinking water is simply taken for granted, as no big deal. Residents may respond to the potential ill effects of the water by boiling it. Over time, they get so used to boiling their water that they don't even think twice about it. Thus they may never really challenge why it is that the water is so bad to begin with. In other places, water provision means something else. It is taken for granted by residents that water is, and ought to be, of good quality. Any negative change to water quality will be met with outrage and concerted public action to clean up the supply. In each of these cases, there may be unsafe drinking water. Sociology can help us discern why different people respond differently to what appears to be much the same problem.

Some questions to ask are:

What is the problem?

To answer this we need to identify the initial problem, such as unsafe drinking water. In order to do this we have to deal with issues of definition and evidence of harm. We have to analyse potentially competing claims as to whether or not the problem exists, and diverse lay and expert opinion on how the problem is interpreted. Does it pose a risk, and if so, to whom, and in what ways? Is the initial problem serious enough in the public's eye to warrant a social response in the form of community action or state intervention?

Why does the problem occur?

To answer this we need to examine the social context, and to investigate the actions of key actors involved. In this instance, we might analyse matters relating to the ownership, control and regulation of drinking water. Who is responsible for water quality? Whose job is it to manage the resource and to whom are they accountable and how?

What are the social dynamics that allow the problem to persist?

To answer this we need to tackle issues pertaining to the shaping of perceptions, interpretation of events, and intervention processes. To explain why unsafe drinking water persists as a problem, we might ask the following subset of questions:

- Is the problem socially constructed as a *social problem* warranting social action; if so, how? (e.g. the emphasis might be on the financial costs of clean-up, or charging for water treatment and use, or making reference to the natural limits of a local environmental resource).
- In what ways is the problem construed from the point of view of *social regulation* and what forms of state and private intervention are mobilised to contain or manage the problem? (e.g. appeal to self-regulation, or regulation premised on the setting of standards, or strong state intervention).
- Is the problem itself to be addressed, or is the focus on how best to avoid, cover up or manage any *risk* associated with the problem? (e.g. signs telling the public not to drink water or to boil it first, installation of water filtration systems).

Regardless of the specific environmental issue or specific social problem, sociological analysis needs to take into account a wide range of concerns.

Implementing 'see, judge, act' in relation to the environment means:

- being cognisant of how environmental issues are socially constructed: *how expertise is mobilised and perceptions influenced by a variety of different actors*.
- identifying the social forces and actors involved in portraying, causing or responding to an environmental issue: *the institutions, people*

and social structures that are associated with a particular trend, event or problem.

- examining how perceptions are influenced by various techniques that affirm or neutralise an issue, how ideas are contested politically and via legal and other means, and how emotions are intertwined in and through public discourses: *the modes of communication and affectation that shape the construction of social problems.*

- investigating how social power is organised in support of particular social interests, in ways that lead to unequal distributions of actual risks and perceived risks: *the ways in which social inequalities are manifest in environmental matters.*

How to comprehend issues and events, and how and on what basis to engage with institutions and groups, are strong thematic currents evident throughout this book. Environmental sociology is, more often than not, about swimming against the tide. In furthering the endeavours of understanding the world, making judgements about it, and acting within it, it is hoped that the book will provide insights into how best to navigate the sometimes murky waters of environmental issues. Good sociology is never far from controversy.

ABOUT THE BOOK

And, of course, this book is very much informed by a sense of 'controversy'. Across a wide diversity of topic areas it is apparent that certain debates and conflicts, specific differences in approach and opinion, and opposing as well as complementary viewpoints come to the fore. The nature, sources and consequences of these controversies provide a useful and interesting way in which to frame environmental issues, and ultimately to understand better the dynamic relationship between society and environment.

The main idea behind the book is to expose the reader to a wide range of intellectual and environmental issues. Indeed, the book's contribution is that it will present ideas and information, and various authors and types of literature, to a wider audience than perhaps has been the case hitherto. The book operates at two levels of exposure. First, it brings together disparate topic areas in a way that allows different types of concerns and issues to be considered in the one volume. These have been grouped under thematic headings: social perspectives, social trends, and social issues. This enables a reasonably cohesive grouping of topics while still maintaining a sense of diversity. Social perspectives allude to ways of seeing the world; social trends refer to patterns and developments; and social issues make reference to specific social problems and dilemmas.

Second, each chapter is structured in such a way as to provide a systematic review of important issues and debates within that particular topic area. The key to environmental sociology is to consider the specifically social dimensions of environmental issues. This requires analysis of the social dynamics that shape and allow certain types of activities pertaining to the environment to take place over time. Each chapter begins by providing a general background to the issues. This is followed by explication of the key debates within the particular topic area. The chapters conclude with a signposting of future directions in the area – analytically, empirically, and with a view to the challenges that lie ahead. The book is meant to describe, to expose, and to excite.

Not only are the topic areas diffuse and variable, but so too are those working on different problems within environmental sociology. As this book demonstrates, one does not have to be a 'sociologist' to do sociology, and sociology itself has many different analytical dimensions. Not surprisingly, we find that much of the debate about environmental issues is intrinsically sociological and certainly multidisciplinary. The boundaries of sociology are not determined by 'who you are', but by 'what you do'. A social science perspective on environmental issues is what unifies the contributions to the present volume.

CONCLUSION – THE BEGINNING

The purpose of this introductory chapter has been to give a rationale for the book, and to provide an analytical context within which the contents might be situated. While the rendition of 'what is environmental sociology' may be somewhat idiosyncratic (reflecting as it does the author's personal interpretation of the discipline), the purpose is not to establish analytical boundaries. Rather, as with the book as a whole, the intention is to open up further conceptual, empirical and action-oriented possibilities.

As a sub-field of sociology, environmental sociology has seen considerable growth in recent years, as much as anything reflecting significant changes in the environment and in public consciousness of these changes. Simultaneously, interest in environmental issues and problems has left no discipline untouched, whether this be economics, political science, geography or law. These developments have also generated extensive cross-disciplinary dialogue and collaboration. In one sense, we are increasingly talking the same language. Yet this language gets ever more complicated and complex.

This book takes its place as a contribution to the diversity of viewpoints, theories and empirical analyses evident across the broad spectrum of social science writing. It is hoped that the debates and controversies described in it will provide markers of where environmental sociology is at today – and where we need to go into the future.

Further Reading

Athanasiou, T. 1996, *Divided Planet: The ecology of rich and poor*, Boston MA: Little, Brown & Co.
Barry, J. 1999, *Environment and Social Theory*, London: Routledge.
Cudworth, E. 2003, *Environment and Society*, London: Routledge.
Franklin, A. 2002, *Nature and Social Theory*, London: Sage.
Hannigan, J. 1995, *Environmental Sociology*, London: Routledge.
Harper, C. 2004, *Environment and Society: Human perspectives on environmental issues*, 3rd edn, New Jersey: Pearson Prentice Hall.
Harvey, D. 1996, *Justice, Nature and the Geography of Difference*, Oxford: Blackwell.
Hay, P. 2002, *Main Currents in Western Environmental Thought*, Sydney: UNSW Press.
Macnaghten, P., and J. Urry 1998, *Contested Natures*, London: Sage.
Wright Mills, C. 1959, *The Sociological Imagination*, New York: Oxford University Press.

CHAPTER ONE

OLD TRADITIONS AND NEW AGES: RELIGIONS AND ENVIRONMENTS

Douglas Ezzy

Surely religion has little to say of significance about the environment? That is a central argument of this chapter. However, it is only half the story, and the opening sentence may not have quite the meaning that you think. It is the Christian tradition and its secularised descendant 'consumerist capitalism' that are the religious traditions that have typically devalued the natural world by ignoring it. This world is of little significance if salvation is primarily in the next world and the key encounter in this world is between an individual's soul and a transcendent deity seen as Other. Similarly, in consumerist capitalism, talk of the rights of trees, fish, or mountains seems strange when human pleasure and wealth are the criteria by which all actions are judged. I argue that at the heart of the current environmental crisis is the relegation of the environment to something of peripheral significance. This relegation derives from the religious traditions of Christianity and consumer capitalism.

Other religious traditions, such as indigenous traditions, Buddhism, and contemporary Paganism, have very different approaches to the natural world. Typically, these traditions regard this earth as important, and do not consider human pleasure and wealth to be adequate justifications for large-scale environmental destruction. The effects of these religious traditions is clearest in their outcomes: they have fostered human societies that live in a largely ecologically sustainable relationship with the forests, rivers, and animals around them.

However, it is too simplistic to blame Christianity for the current environmental crisis and point to other religious traditions as solutions. Gottlieb (1996a: 9) argues: 'religions have been neither simple agents of environment domination nor unmixed repositories of ecological wisdom. In complex and variable ways, they have been both.' Indigenous traditions have been involved in ecologically destructive activities (Flannery 1994). Some Christians have advocated a more sensitive approach to the environment (Tucker 2003). Further, the social and cultural formation of contemporary consumerist capitalism has a quasi-religious dynamism all of its own (Greider 1997; Loy 1997) that has played a central role in environmental destruction.

Nonetheless, Lynn White (1967: 1204) was surely correct when he argued in 1967 that: 'More science and more technology are not going to get us out of the present ecologic crisis until we find a new religion, or rethink our old one.' At the heart of the contemporary ecological crisis is a theological, and sociological, problem. The destruction of huge sections of the world's ecosystems is a product of a culture imbued with theologically derived beliefs about the relationship of humans to the non-human world.

BACKGROUND TO THE ISSUES

> What they [environmentalists] want from religion happens to be, many would say, the most decisive ingredient in any effective environmental ethic: the ability to move from an anthropocentric to a biocentric understanding of the world and our human place in that world. Environmentalists have long recognized this shift as essential; recently, many are also recognizing – some with consternation, others with hope – that this shift is really 'a religious question'. (Knitter 2000: 377)

Anthropocentrism is a way of viewing the world, and choosing how to act in the world, in which human welfare and concerns are the final arbiter of what should or should not be done. In its strongest form anthropocentrism argues that the natural world only has value when it becomes a product for human consumption (Hay 2002). Another, slightly less arrogant, form of anthropocentrism argues that the non-human world is valuable when it is instrumental to human purposes (Hay 2002). Anthropocentrism is characteristic of Christianity and the Western capitalist worldview.

In contrast *ecocentrism*, also referred to as *biocentrism*, is concerned with sustaining the whole of an ecosystem. Humans are envisaged as one of a variety of beings with value in an ecosystem. Ecocentrism is common among indigenous societies, some forms of Buddhism, contemporary neo-Paganism, and the deep ecologists. It is often constructed as diametrically opposed to anthropocentrism.

Joanna Macy (1991: 32) describes *deep ecology* as an awareness of how humans are 'interwoven threads in the intricate tapestry of life'. From this ecological perspective all life is part of various open systems that are self-organising. Deep ecology and ecocentrism argue that all beings, not just humans, have rights.

The idea that beings other than humans may not only have rights, but also be ontologically, and perhaps even metaphysically, integral parts of what it means to be human is not an easy idea for many people raised within the context of a Western philosophical and scientific worldview. Even if this idea remains marginal to much contemporary thought, it is increasingly acceptable within academic discussions of ecology, sociology, theology, and religious studies. It is one of the central points of debate in contemporary studies of religion and ecology. It also reflects a much broader debate about the nature of what it means to be human and what constitutes ethical action.

Much of Western religious thought, and the philosophical tradition that has developed alongside it, emphasises transcendent sources of morality, divine commandments, and logical categories for understanding. Following Descartes' philosophy, it also begins with isolated individuals, building the world up and out from the reality and rights of individuals. This is the dominant anthropocentric individualism of Christianity and consumerist capitalism.

In contrast, the ecocentrism of the deep ecologists has many similarities with the hermeneutical theory of Gadamer, Charles Taylor and to a lesser extent, Bauman (Ezzy 1998). In this communitarian tradition, the starting point is not individuals, but relationships: 'all real living is meeting' (Buber 1958: 11). This is a radically different way of understanding the human condition that does not proceed from the individual out to relationships, but begins with relationships, and views the individual as arising in and out of these relationships. Buber and the deep ecologists include humans in these relationships along with trees and other aspects of nature.

It is important to understand that I am making sociological and historical points, not theological. That is to say, I am not making a theological argument about what Christians *should* believe. Rather, I am describing sociologically and historically what most of the people who have called themselves Christians have believed for approximately the last 500 years. Most Christians have not defended the rights of nature. 'Instead people used Scripture to justify the exploitation of nature in the same way that the defenders of slavery used it to justify ownership and exploitation of certain classes of humans' (Nash 1989: 91).

In contrast, indigenous traditions often saw humans as one part of a broader society that included other non-human beings. Humans had an ethical responsibility for these other beings:

> Central to most Indian religions and ethical systems was the idea that humans and other forms of life constituted a single society. Indians regarded bears, for example, as the bear *people*. Plants were also people. Salmon constituted a nation comparable in stature and rights to human nations. A complex of rituals and ceremonies reinforced the familiar bonds between Indians and their environment. (Nash 1989: 102)

The focus in this chapter on Christianity is not a form of ethnocentrism. Commentary on the ecological orientations of other world religions is important. However, it is Christianity and its secularised descendant of consumer capitalism that are the primary influence on the current ecologically destructive practices that threaten the world's ecosystems. For example, Dwivedi (1996) points out that it may be too simplistic to entirely blame India's ecological crisis on the culture and practice of Christian British imperialism that dominated India for 700 years. The influence of Islam and internal transformations in Hinduism must also be taken into account. However, the influence of Christianity and Western secular institutions and values 'greatly inhibited . . . the ancient educational system [of Hinduism] which taught respect for nature and reasons for its preservation' (Dwivedi 1996: 161).

KEY DEBATES

An environmental crisis?

Is there an environmental crisis? That there is an environmental crisis, and that it is very serious, is an almost taken-for-granted fact by all authors in the various anthologies available on religion and the environment (Gottlieb 1996a; Hessel & Ruether 2000; Grim 2001b; Foltz 2003a). This may reflect the assumptions and beliefs of the editors of these collections. More probably, however, it reflects the social distribution of understandings about the environment.

The 1995 Report to the Intergovernmental Panel on Climate Change was passed by 2500 scientists with no dissenting or minority report (McFague 2000). This report warned of the potentially devastating ecological, economic and social consequences of climate change and recommended immediate action. Big business, however, still tries to create a public perception that there is disagreement among scientists over the reality and consequences of global climate change. McFague (2000: 39) observes: 'Climate change challenges the fossil fuel industry, as well as America's love affair with the car. Hence, denial and resistance are high.'

The denial by big business of the current ecological crisis has deeper cultural roots than short or medium-term economic interests – there is no such thing as long-term economic planning as almost no business plans beyond fifteen years (Weizsäcker et al. 1977). Specifically, as Weiskel puts it: 'Without any exaggeration . . . it is fair to say that in practical terms the most pervasive form of this religiously held belief in our day is that of growthism founded upon a doctrine of techno-scientific salvation' (Weiskel quoted in Knitter 2000: 371). The leaders of big business seem convinced, by and large, that scientific advances and new technology will solve any environmental problems that may arise in the future (Ezzy 2001). This is why big business still refuses to be concerned about environmental issues, as evidenced by the fact that no piece of environmental legislation in the United States has ever been supported by the big business lobby (Barry 1999).

The role of Christianity

What is the contribution of Christianity to the Western attitude towards the environment? Although earlier forms of Christianity retained some interest in the natural world, most commentators agree with Johnson's (2000: 4) assessment that 'for the last five hundred years the religious value of the earth has not been a subject of theology, preaching, or religious education'. If anything, the natural world, including the human body, was seen as dangerous and a thing to be disciplined. In the modern period Christianity saw nature as something that people were rescued from. Haught (1996: 270) argues that most Christians continue to see this world as little more than a 'soul school', a background to the real journey of salvation in the next world.

The clearest illustration of the anthropocentric individualism of Christianity comes from the Protestant Reformation's emphasis on 'faith alone'. Protestantism emphasised that individuals had to have a personal encounter with God. Through Christ alone and through cognitive understanding of scripture the individual came to an individual relationship with God (Johnson 2000). As a consequence Christianity has ignored ecological concerns. McFague (2000: 39), who is a Christian, puts it succinctly: 'For the past several hundred years, Christians have not had a practice of loving nature; we have not practised justice toward nature, nor cared for it. We have lacked a well-informed, respectful, unsentimental concern for nature similar to that which we have tried to develop toward other human beings.'

However, as I have already indicated, Christianity on its own cannot be blamed for the current ecological crisis. Anthropocentrism is also deeply rooted in the social and economic organisation of contemporary Western societies. Cowdin (2000) argues that the main sources of the current ecological crisis lie with philosophical and scientific developments during the 15th and 16th centuries, and not with Christianity. Descartes' understanding of the self separated the mind from the body. Newtonian physics turned the material world into an inert mechanical process devoid of spirit or value. 'Bacon's manipulative scientific domination of nature . . . combined to form the uniquely modern view of nonhuman nature as simply a valueless field to be exploited for human benefit' (Cowdin 2000: 265). Similarly, Kantian philosophy divorced the knowing active subject from an external passive nature.

Alongside these philosophical changes were changes in social life and the economy that worked hand in hand with the new philosophies. Industrial capitalism moved people off the land and into cities, breaking up traditional communities and traditional relationships with the cycles of the land. Berry (2000: 130) argues that one major source of humans' loss of intimacy with the natural world 'occurred at the end of the nineteenth century when we abandoned our role in an ever-renewing organic agricultural economy in favor of an industrial non-renewing extractive economy'.

On the other hand Hallman (2000: 458) observes that the Protestant Reformation played a key role in the development and spread of the scientific worldview and philosophy: 'In emphasizing the distinction between the Creator and the creature, the Reformers desacralized the natural world. This made it possible for the budding sciences to study and manipulate it.'

On balance it is probably accurate to say that both the Reformation and the Scientific Revolution in the 16th and 17th centuries represent key transitions in the way Westerners have understood their relationship to nature. 'For Calvinism, nature was totally depraved . . . Saving knowledge of God descends from on high, beyond nature, in the revealed World available only in Scripture, as preached by the Reformers' (Ruether 1996: 328). Similarly, 'in Cartesian dualism and Newtonian physics nature becomes matter in motion, dead stuff moving obediently, according to mathematical laws knowable to a new male elite of scientists' (Ruether 1996: 328). Both these movements made an anthropocentric world possible in which nature could be exploited and destroyed with few moral concerns.

The centrality of consumerist capitalism to contemporary Western lives makes it extremely difficult to change. As Cobb (2000: 497) observes: 'Even Christians who cease to be anthropocentric and dualistic, and recognize the intrinsic value of nonhuman creatures, often remain committed to the technology that drives the progressive deterioration of the natural world.' The majority of Christians in Europe and the United States are blind to how closely bound up their version of Christianity is with contemporary consumer capitalism (Robra 2000). Christians in these countries often speak disparagingly of syncretism between Christianity and other cultural traditions in countries from the South, but fail to see precisely the same processes in operation in their own Christian culture. Or put more bluntly, 'Christians share with [Western] society as a whole the short-term selfishness that we deplore' (Cobb 2000: 497).

Is it possible to develop an ecologically sensitive Christianity? Although there are other developments (see Hessel & Ruether 2000), eco-feminism is probably the most widely recognised. 'The basic assumption of eco-feminist theology . . . is that the dualism of soul and body must be rejected, as well as the assumptions of the priority and controlling role of male-identified mind over female-identified body' (Ruether 2000). There is a substantial eco-feminist literature that draws on nature imagery in the Bible and that reclaims an ecocentric model of the spirit in nature. These new theologies lead to a more this-worldly orientation.

While eco-feminists and eco-theologians work hard to theorise and theologise alternatives, these are not widely accepted in Christian cultures. Berling (2003) makes precisely this point in her response to Mary Tucker's call for religions to become more ecologically oriented. It may be true that there is a pressing need for religions to focus on ecological concerns. However, Berling (2003: 59) admits, 'I have a hard time envisioning how this issue will move front and centre in the thinking of any of the religions: that is to say, we may worry about our "primary" issues within each of the religions while the planet dies around us.' Further, a number of Christian theologians have condemned the this-worldly orientations of the eco-feminists and eco-theologians as Paganism (Jones 1999). Clearly there is considerable resistance within the Christian tradition for a move away from the Protestant emphasis on salvation to an ecocentric theology.

In contrast to indigenous understandings of forests and animals as having voices and rights, materialist modernist science silenced these voices. This is not simply a philosophical and metaphysical silencing. The unparalleled rate of species extinctions that is currently occurring (Brown & Flavin 1999) is a very concrete and literal silencing of the animals and plants. As Walsh and associates (1996: 423) write, 'Rivers are muted when they are dammed; prairies are silenced when they are stripped for coal; mountains become torpid when they are logged.'

Indigenous responses to environmental problems

Indigenous understandings of their ecological environment typically stand in sharp contrast to a Western worldview. Indigenous societies tend to see humans as one part of a larger community that includes animals, plants, and local geographical features. Further, indigenous societies tend to interpret these relationships in mythological terms. Plants and animals are viewed as spiritual beings who should be treated with respect, and who have power to shape the lives and prosperity of humans. As Grim (2001b: lv) so elegantly puts it, indigenous cultures have 'imagined themselves more intimately into their worlds'. Indigenous cultures often represent a sophisticated ecological understanding of local ecosystems and how to manage them sustainably.

It is important to underline that indigenous worldviews are no less legitimate than the mythology of Western scientific humanism. Garfield (2002) has shown that it requires a particular form of ethnocentric post-colonial racism to deny that the philosophical practices of Buddhists are indeed philosophy. Similarly, the presumption that indigenous worldviews are by definition inferior, and that the European Western scientific worldview is in some fundamental way superior represents ethnocentric hubris. I am not arguing that all indigenous worldviews are always superior. Rather, I argue that indigenous worldviews and ecological practices should be given the same respect and critical scrutiny as the Western scientific worldview and associated practices of economic development.

Virtually all indigenous societies around the world are shaped by a history of conquest and exploitation. Cultural imperialism, through the efforts of Christian missionaries, and economic exploitation, encouraged by multinationals, are the two primary influences. Indigenous societies are typically numerically small, politically weak, economically poor, and with a history of cultural degradation through the impact of Christian or Islamic missionaries. Surrounding societies are typically numerically large, politically strong and economically wealthy, and with a deeply held view that indigenous cultures are intrinsically inferior to their own (Greaves 2001).

Fried (2001), for example, provides an extended account of the Bentian indigenous society in Borneo. Christian missionaries attempted to destroy the longhouses, which were the cultural centres of Bentian community, and stop the celebration of agricultural and forest-based indigenous Kaharingan religious ceremonies. Further, a logging company sought Christian blessing for a logging road, the construction of which was prevented by ancestral indigenous demons who were haunting the area.

> Opposition to Bentian forms of worship moved from the ideological basis of the missionaries to the commercial needs of the logging companies and their desire to manipulate Christian and Kaharingan rituals alike, in order to maximize profits. Most of these attempts, in one way or another, made life more complicated and stressful for the Bentian, but – as they continued to battle logging and plantation company encroachment on their forests and continued, more or less, to celebrate the new harvest and old weekly *belian* healing ceremonies – the strenuous efforts made to influence Bentian cosmology did not necessarily achieve the goals of either church or logging company. (Fried 2001: 97)

Similarly, Sharma's (1996) speech to the Earth Summit in Rio, Brazil, in 1992, provides an incisive account of an indigenous Indian community attempting to resist state-sponsored logging of the forests that have historically sustained their community. Sharma makes a sustained argument that the logic of contemporary global capitalism serves to impoverish, rather than enrich, indigenous communities. He argues that the pleasure of consumerist delicacies does not justify the destruction of the foundations of the livelihood of his local community. 'Look again at your world. The community has already been sacrificed on the altar of productivity' (Sharma 1996: 561). Sharma's point is not that all development is wrong, or that progress is not possible, but that development as it is currently implemented typically results in the impoverishment of indigenous communities: 'What we insist on is that development must have a human face, or else it is tantamount to destruction' (Sharma 1996: 564).

Western cultures have typically assumed that indigenous cultures are 'backward' or in some fundamental way inferior. This assumption has been driven by the resistance in indigenous cultures to economic, technological, and agricultural developments. This has often led Westerners to assume that it is in the best interests of indigenous societies to abandon their indigenous religious traditions, which are often animistic and antithetical to Western individualism. This view remains dominant in many Western organisations, including governments and global multinationals. However, it has some serious flaws. In particular, it is clear that in most cases in which 'development' has been forced on indigenous communities it has resulted in significant impoverishment of these communities (Grim 2001b).

Indigenous peoples have typically not separated cosmology from economy. Their mythologies and religious practices were, and in some cases still are, integral to sustainable agricultural and ecosystem management strategies. As such, the economically driven destruction of the local habitats of indigenous communities is fundamentally a religious issue. To see it in any other way is to adopt the imperialistic exploitative worldview of Western capitalism which is at the heart of the current ecological crisis.

The globalisation of contemporary consumerist capitalism has placed indigenous societies under enormous pressure to assimilate into mainstream Western cultures and to make their lands available for the exploitatively destructive practices of forestry, mineral extraction, fishing, or tourism. According to this view the destructive practices of Western-style development are simply normal and rational. Against this is a more sophisticated, and less ideologically driven, understanding that 'indigenous peoples have alternative development models that value homelands differently . . . [that involve the] use of those lands [for] living for food, habitat, and trade'. By so doing, 'they embody alternative models of sustainable life' (Grim 2001b: xl).

Buddhism and Islam

Buddhism is often described as one of the more environmentally oriented religious traditions. Buddhism has been drawn on by a number of ecological thinkers to provide intellectual and practical resources to counter the environmental destruction of modernity (Snyder 1990; Macy 1991). In the Chinese version of Buddhism there is a tradition of a close connection between humans and nature where nature is seen as healing and supportive. According to one interpreter the Chinese Buddhist tradition teaches that:

> Everything has a Buddha-nature . . . [and] every person has a Buddha-nature, but what was of such importance to the Chinese was the teaching that insentient objects also have it. The rocks, trees, lotuses, streams, mountains – all have Buddha-nature. Therefore, one's mind, which has Buddha-nature as its essence, shares a common aspect with every part of insentient nature, which also possess this same Buddha-nature. (Lancaster 1997: 13)

Compared to Christianity, Buddhism has a different understanding of how humans should respond to the dangers and difficulties of life. Rather than seeking to escape these dangers by destroying them or overcoming them, 'in the Buddhist texts and teachings we hear the hard truth that none of these perceived dangers will remain unchanged or permanent, and we must learn how to survive in a natural state of constant change' (Lancaster 1997: 13). In contrast to Christianity's attempt to transcend this world and death, some traditions of Buddhism seek to live with the flux of life and death in the world.

However, the practice of the Buddhist traditions varies considerably in their understanding of nature. Buddhism is a complex and culturally diverse religious tradition. For example, the Japanese Buddhist philosopher Nishida Kitaro developed a doctrine of 'pure nondual experience' that embodies a vision of spirituality as nature-transcending subjectivity (Harris 1997: 385). This approach does not lead to a close connection to nature.

Muslim intellectuals are also beginning to examine the environmental crisis. The Qur'an contains many references to nature, encouraging the careful use of resources such as air, water, and agricultural land. 'Yet the articulation of an Islamic environmental ethic in contemporary terms . . . is all quite new' (Foltz 2003b: xxxviii).

Muslim scholars have emphasised the social justice aspect of the consequences of environmental degradation, as many Muslims live in poorer countries. Environmental harm shapes the lives of the poor more than the rich, who are able to insulate themselves from its effects (Foltz 2003b: xxxix). Both Islamic and indigenous thinkers emphasise that ecology cannot be separated from economy, and that to do so represents a particularly ethnocentric view of the world that has developed out of the Christian West.

While Muslim thinkers typically recognise the serious nature of environmental degradation, they have argued that this is not the primary problem. Rather, environmental degradation is a symptom of a deeper moral and religious evil in which contemporary societies have turned away from God. 'A just society, one in which humans relate to each other and to God as they should, will be one in which environmental problems simply will not exist' (Foltz 2003b: xxxix).

This argument has more plausibility than it may at first seem. It points to the centrality of consumer capitalism to current environmental problems. Dutton (2003) argues that environmental destruction is largely a consequence of greedy profit-driven exploitation of resources to the detriment of human communities. 'What people have not been told is that it is usury that underpins this whole economy . . . and usury . . . is totally forbidden in Islam' (Dutton 2003: 331). Usury is the practice of charging interest on monetary loans.

Ecological degradation is often described in Western mass media as solely a technological or scientific problem. This is a particularly ethnocentric view. There is still room for debate about the solutions recommended by Muslims, eco-feminists, Buddhist and indigenous thinkers. However, they all agree that the source of the problem lies as much in cultural and religious understandings as it does in technological problems.

The New Age and contemporary Paganism

The New Age is not really a religious movement so much as a general trend in spirituality in Western societies. The attitude towards nature of most of these New Age spiritualities mirrors the Western attitude towards nature. They tend towards an anthropocentric individualism. New Age spirituality has typically involved transforming the lives of individual people through a search for inner harmony and bliss or empowering prosperity (Heelas 1996). Some New Age spiritualities that seek change through counter-cultural living have developed a different, more ecocentric approach to ecological problems (Sutcliffe 2003).

The contemporary Witchcraft movement is one part of a more general explosion of Pagan religions in the Western world (Hume 1997). Although the numbers are small, this religious tradition has become very attractive to young people, in part because of its explicitly nature-based spirituality. Witchcraft involves the celebration of religious festivals dictated by the seasons of the sun and the moon. It also treats forests, animals, mountains and rivers as sacred. Some Witchcraft traditions are explicitly ecologically activist, and involved in protests against nuclear power and the destruction of forests (Harvey 1997). However, other Witchcraft traditions are not environmentally activist and reflect a New Age focus on self-transformation through spiritual practices (Greenwood 2000).

The key debate here is whether spiritual practices that transform self-understandings are an adequate response to environmental degradation. The focus on the self of New Age spiritualities and Witchcraft could be seen as simply managing the individual consequences of the environmental crisis rather than addressing its cultural and structural sources. Perhaps New Age and Pagan spirituality is another form of short-term selfishness (Letcher 2000). I have argued elsewhere that the more Witchcraft becomes incorporated into the goal of making money the more it tends to focus on selfish individualism and lose its potential for social transformation (Ezzy 2003).

Religion, the environment, and ethics

The most important debate is whether the current rapid and severe degradation of the environment matters. The leaders of big business, politicians, and most Christians seem to think that environmental degradation is not a problem that demands immediate action. In contrast, scientists and some religious leaders are in agreement that there is a serious problem and something needs to be done about it urgently.

In 1992 over sixteen hundred senior scientists from seventy-one countries signed a document titled 'World Scientists' Warning to Humanity'. The document begins: 'Human activities inflict harsh and often irreversible damage on the environment and on critical resources. If not checked, many of our current practices put at serious risk the future that we wish for human society and the plant and animal kingdoms' (Suzuki 1997: 4). The release of this document was not covered at all in the North American mass media. Instead, one of the main North American newspapers, in one of the most Christian countries in the world, had a large photograph of cars forming an image of Mickey Mouse (Suzuki 1997: 5). Do Christianity and consumerist capitalism still have nothing significant to contribute to the solution to environmental degradation?

Lynn White (1967: 1206) suggested in 1967 that 'we are not going to get us out of the present ecologic crisis until we find a new religion, or rethink our old one'. The Christian tradition has a strong history of saving and valuing human life. Similarly, consumerist capitalism is focused on enriching the pleasure of the lives of the wealthy. However, these are of questionable relevance to a world that is overpopulated with humans who consume too much of the earth's resources. It could be argued that Christianity and consumerist capitalism are no longer useful, but rather extremely harmful religious traditions.

As I have already argued, it is too simplistic to dismiss Christianity. Religion has historically played a central role in democracy. This is clearest in its role in limiting self-interest and requiring society to take care of the poor and less well off. Schools, hospitals, aged care facilities, and a variety of other welfare organisations, all began in Western societies under the auspices of various churches. Cobb (2000) argues that the failure, until recently, of Christianity to address the social and economic structures that lead to ecological degradation is not an inherent result of the Christian faith. Instead he believes, along with a small proportion of other Christian thinkers, that the Christian tradition offers important resources for addressing current ecological problems. It remains to be seen whether the ideas of this vanguard of Christian thinkers are translated into real changes in practice.

Much of the ecological literature, and some Christian writings about the 'end time', seem to suggest the earth is heading for ecological catastrophe (Ezzy 2001). However, other ecological activists and religious leaders have tried to be more positive and to imagine alternative ways of living that are ecologically sustainable. Indigenous communities with nature-based mythologies may provide a much more sophisticated way of living with local ecosystems. Buddhist traditions that emphasise the shared Buddha-nature of humans and other beings suggest a different way of understanding the relationship of humans to the rest of nature. Christians and Muslims have pointed out that social justice is an integral part of a sophisticated response to ecological degradation. Deep ecologists and some contemporary Pagans have begun to develop a mythology that reimagines non-human beings into an ecocentric world in which these beings have value in and of their own right alongside humans.

FUTURE DIRECTIONS

At the heart of the current massive environmental degradation is a culturally and economically entrenched Western practice of ignoring the environmental consequences of human actions. Western culture and Western economic practices are also powerfully subversive of any alternative culture and practices that challenge their dominance. From this perspective it is difficult to see how an effective social transformation is possible. However, as the classical sociologist Max Weber argued, religion provides one of the few sources of charismatic authority that can motivate resistance to the iron cage of capitalist rationality (Weber 1947). As such, the key debates of the future will focus around the tensions between, on the one hand, religiously inspired evaluations of the environment that will be increasingly ecocentric and, on the other hand, consumerist capitalism that is integrally anthropocentric, arrogant, aggressive, and powerful.

Discussion Questions

1. Why is it important that religious traditions address ecological degradation?
2. What is the difference between a theological analysis and a sociological analysis of the orientations of religions to nature?
3. Compare the influence of Christianity and capitalism on current environmental practices.
4. What is eco-feminism?
5. What are the key characteristics of indigenous approaches to ecological management?
6. Explain how indigenous traditions do not separate spirituality from economy.

7. Does Buddhism provide intellectual and practical resources that are more ecologically sensitive?
8. How have Muslim thinkers responded to ecological concerns?
9. Are the New Age and contemporary Paganism just a spiritual form of selfish individualism?
10. Big business is largely unconcerned about ecological degradation. How is this related to religious understandings of ecology?

Glossary

Anthropocentrism: the view that humans are the centre of the universe, and that things only become valuable when they can be used by humans.

Biocentric or ecocentrism: the view that humans are just one of a number of beings on the earth and that all have rights and values.

Consumer capitalism: refers to contemporary societies in which purchasing and consuming are central to most people's lives. For example, in consumer capitalism a person's status is more influenced by what they wear, eat, or drive, than by what they consume, what sort of job they have, or their family connections.

Deep ecology: the view that sees humans as one part of an intricate web of life. This is linked to the idea that plants, animals, mountains, and rivers have rights of their own and are not merely there for human pleasure.

Eco-feminist theology: a recent development within Christianity that seeks to reclaim biblical images of women and of nature, seeking to move beyond the patriarchal and otherworldly focus of traditional Christianity.

Paganism: a variety of contemporary religious traditions including Witchcraft, Druidry and Heathenism that focus on celebrating the natural world. This typically involves religious festivals linked to the seasons of the Sun, such as solstices and equinoxes, and the seasons of the Moon, such as full and dark moons. These traditions usually focus on experience rather than belief, and this-worldly pleasure rather than salvation in the other world.

Religion: a set of symbols and beliefs embodied in a social institution, such as a church, providing a set of ritual practices for maintaining contact between this world and a system of transcendent meaning and experience (Hanegraaf 1999).

Spirituality: an individual's set of symbols and beliefs that provide that individual with rituals to maintain contact between this world and a system of transcendent meaning and experience (Hanegraaf 1999).

References

Barry, J. 1999, *Environment and Social Theory*, New York: Routledge.
Berling, J. 2003, 'Commentary'. In M.E. Tucker, *Worldly Wonder: Religions enter their ecological phase*, Chicago: Open Court.

Berry, T. 2000, 'Christianity's role in the earth project'. In Hessel and Ruether, *Christianity and Ecology*.

Brown, L., and C. Flavin (eds) 1999, *State of the World 1999*, London: Earthscan.

Buber, M. 1958, *I and Thou*, transl. R. Smith, New York: Collier Books.

Cobb, J. 2000, 'Christianity, economics, and ecology'. In Hessel and Ruether, *Christianity and Ecology*.

Cowdin, D. 2000, 'The moral status of otherkind in Christian ethics'. In Hessel and Ruether, *Christianity and Ecology*.

Dutton Y. 2003, 'The Environmental crisis of our time: a Muslim response'. In R. Foltz, F. Denny and A. Baharuddin (eds) *Islam and Ecology*, Cambridge MA: Harvard University Press.

Dwivedi, O. 1996, 'Satyagraha for conservation: awakening the spirit of Hinduism'. In Gottlieb, *This Sacred Earth*.

Ezzy, D. 1998, 'Theorizing narrative-identity: symbolic interactionism and hermeneutics', *Sociological Quarterly* 39(2): 239–52.

Ezzy, D. 2001, 'Reading for the plot, and not hearing the story: ecological tragedy and heroic capitalism'. In A. Mills and J. Smith (eds) *Utter Silence*, New York: Peter Lang.

Ezzy, D. 2003, 'New age witchcraft?' *Culture and Religion* 4(1): 47–66.

Flannery, T. 1994, *The Future Eaters*, Sydney: Reed New Holland.

Foltz, R. 2003a, *Worldviews, Religion, and the Environment*, Melbourne: Thomson.

Foltz, R. 2003b, 'Introduction'. In R. Foltz, F. Denny and A. Baharuddin (eds) *Islam and Ecology*, Cambridge MA: Harvard University Press.

Fox, M. 1991, *Creation Spirituality*, San Francisco CA: Harper.

Fried, S. 2001, 'Shoot the horse get the rider: religion and forest politics in Bentian Borneo'. In J. Grim (ed.) *Indigenous Traditions and Ecology: The interbeing of cosmology and community*, Cambridge MA: Harvard University Press.

Garfield, J. 2002, 'Philosophy, religion, and the hermeneutic imperative'. In J. Malpas, U. Arnswald and J. Kerscher (eds) *Gadamer's Century*, Cambridge MA: MIT Press.

Gottlieb, R. 1996a, 'Religion in an Age of Environmental Crisis'. In Gottlieb, *This Sacred Earth*.

Gottlieb, R. (ed.) 1996b, *This Sacred Earth: Religion, nature, environment*, New York: Routledge.

Greaves, T. 2001, 'Contextualizing the Environmental Struggle'. In Grim, *Indigenous Traditions and Ecology*.

Greenwood, S. 2000, *Magic, Witchcraft and the Otherworld*, Oxford: Berg.

Greider, W. 1997, *One World, Ready or Not: The manic logic of global capitalism*, Harmondsworth, UK: Penguin.

Grim, J. (ed.) 2001a, *Indigenous Traditions and Ecology: The interbeing of cosmology and community*, Cambridge MA: Harvard University Press

Grim, J. 2001b, 'Introduction'. In Grim, *Indigenous Traditions and Ecology*.

Hallman, D. 2000, 'Climate change: ethics, justice and sustainable community'. In Hessel and Ruether, *Christianity and Ecology*.

Hanegraaf, W. 1999, 'New age spiritualities as secular religion', *Social Compass* 46(2): 145–60.

Harris, I. 1997, 'Buddhism and the discourse of environmental concern'. In M. Tucker and D. Williams (eds) *Buddhism and Ecology*, Cambridge MA: Harvard University Press.

Harvey, G. 1997, *Listening People, Speaking Earth: Contemporary paganism*, Adelaide: Wakefield Press.

Haught, J. 1996, '"Christianity and Ecology" from *The Promise of Nature*'. In Gottlieb, *This Sacred Earth*.

Hay, P. 2002, *Main Currents in Western Environmental Thought*, Sydney: UNSW Press.

Heelas, P. 1996, *The New Age Movement*, Oxford: Blackwell.

Hessel, D., and R. Ruether (eds) 2000, *Christianity and Ecology*, Cambridge MA: Harvard University Press.

Hume, L. 1997, *Witchcraft and Paganism in Australia*, Melbourne University Press.

Johnson, E. 2000, 'Losing and finding creation in the Christian tradition'. In Hessel and Ruether, *Christianity and Ecology*.

Jones, P. 1999, *Gospel Truth, Pagan Lies*, Enumclaw, WA: Winepress Publishing.

Knitter, P. 2000, 'Deep ecumenicity versus incommensurability: finding common ground on a common earth'. In Hessel and Ruether, *Christianity and Ecology*.

Lancaster, L. 1997, 'Buddhism and ecology: collective cultural perceptions'. In M. Tucker and D. Williams (eds) *Buddhism and Ecology*, Cambridge MA: Harvard University Press.

Letcher, A. 2000, '"Virtual paganism" or direct action?' *Diskus* 6, http://www.uni-marburg.de/religionswissenschaft/journal/diskus.

Loy, D. 1997, 'The religion of the market', *Journal of the American Academy of Religion* 65(2): 275–89.

Macy, J. 1991, *World as Lover, World as Self*, Berkeley CA: Parallax Press.

McFague, S. 2000, 'An ecological Christology: does Christianity have it?' In Hessel and Ruether, *Christianity and Ecology*.

Namunu, S. 2001, 'Melanesian religion, ecology and modernization in Papua New Guinea'. In Grim, *Indigenous Traditions and Ecology*.

Nash, R. 1989, *The Rights of Nature: A history of environmental ethics*, Madison WI: University of Wisconsin Press.

Posey, D. 2001, 'Intellectual Property Rights and the Sacred Balance'. In Grim, *Indigenous Traditions and Ecology*.

Robra, M. 2000, 'Response to Marthinus Daneel'. In Hessel and Ruether, *Christianity and Ecology*.

Ruether, R.R. 1996, 'Ecofeminism: Symbolic and social connections of the oppression of women and the domination of nature'. In Gottlieb, *This Sacred Earth*.

Ruether, R.R. 2000, 'Conclusion: eco-justice at the centre of the Church's Mission'. In Hessel and Ruether, *Christianity and Ecology*.

Sharma, B. 1996, 'On sustainability'. In Gottlieb, *This Sacred Earth*.

Snyder, G. 1990, *The Practice of the Wild*, San Francisco CA: North Point Press.

Sutcliffe, S. 2003, *Children of the New Age*, New York: Routledge.

Suzuki, D. 1997, *This Sacred Balance*, Sydney: Allen & Unwin.

Tucker, Mary 2003, *Worldly Wonder: Religions enter their ecological phase*, Chicago: Open Court.

Walsh, B., M. Karsh and N. Ansell 1996, 'Trees, forestry, and the responsiveness of creation'. In Gottlieb, *This Sacred Earth*.

Weber, M. 1947, *Max Weber: The theory of social and economic organization*, transl. T. Parsons, New York: Free Press.

Weizsäcker, E., A. Lovins and L. Lovins 1997, *Factor 4*, London: Earthscan.

White, L. 1967, 'The historical roots of our ecological crisis', Science, 155(3767): 1203–7.

CHAPTER TWO

SOCIAL NATURE
The Environmental Challenge to Mainstream
Social Theory

Stewart Lockie

Until recently, sociological theory has had little to say about nature or
the environment. Reflecting its origins in the social transformations of
the Industrial and French Revolutions, sociological theory adopted the
modernist ideology that through the application of labour and creativity humankind could emancipate itself from the animalistic fight for
survival characteristic of other species (Latour 1993). The blowtorch
of sociological analysis was turned towards a host of institutions ranging from religion and art to education and class relations in order to
illuminate and confront relationships of domination and control. But
nature, and the multitude of organisms, substances and patterns that
comprise it, were taken for granted as passive participants in this great
human drama. Under the sway of modern science and technology, the
environment had lost its mystical and autonomous status and become
a blank canvas onto which human aspirations and projects could be
painted.

With hindsight, the naivety, if not arrogance, of modernist ideology
seems obvious. Social struggles over genetic engineering, nuclear proliferation, toxic waste, food safety, environmental justice, and so on, blur
the false distinction between social and environmental issues (Latour
1993). These struggles refuse to comply with neat divisions of labour
between the social and the natural sciences and force environmental movements to extend their concerns beyond traditional issues of
'wilderness' preservation and species conservation (see Pepper 1984) to
include indigenous peoples' rights, rural livelihoods, public health and
the use of urban space (Lockie 2004). Social theory that cannot find a
place for the non-human organisms, substances and patterns of nature
is social theory that is inadequate for understanding key dimensions of
our contemporary world.

Bringing nature back into social theory requires us to do much more than simply apply existing sociological concepts to a new topic – 'the environment'. Instead, it requires us to rethink many of the basic assumptions on which sociological theory rests and in this way represents a major theoretical challenge to mainstream sociology. To many, this will sound a big, if not ludicrous, claim. How can a minor sub-discipline such as environmental sociology mount such a significant challenge to the core assumptions of its parent discipline? The answer to this question lies partly in the allies environmental sociology finds in a small number of other recently emergent sub-disciplines that also force us to revisit our understanding of sociological theory and method – in particular the sociology of the body (see Turner 1996) and branches of feminism such as eco-feminism. The answer can also be found in existing sociological theory.

This chapter reviews some of the most prominent attempts to bring nature into social theory. In this sense it does not offer a comprehensive overview of all relevant social theory but is structured around four major perspectives that correspond with the major theoretical controversies characteristic of environmental social theory. It argues that the most convincing of these perspectives are those that abandon dichotomies (or polemical, dualistic ways of thinking) between the social and the natural. Such approaches have a great deal to offer mainstream sociology because, at the same time that they reformulate basic sociological concepts such as power and agency, they incorporate a range of conceptual and methodological tools with which to examine the ways in which people, technologies, plants, animals and other entities interact to shape 'social nature'.

BACKGROUND TO THE ISSUES

Sociological theories may be understood as the main schools of thought within the discipline of sociology (Jary & Jary 1991). Theoretical perspectives within sociology provide researchers and students with guidance as to what sorts of problems and questions may be in need of investigation, the concepts and hypotheses that might guide that investigation or explain its outcomes, and the methods appropriate to undertake it. While there are diverse theoretical perspectives within sociology, they have in common an assumption that the root cause of all social phenomena is social relationships. Durkheim (1966) expressed this with the aphorism that social facts must always be explained with other social facts. C. Wright Mills (1959) expressed it in a slightly different manner in suggesting that the 'sociological imagination' had a particular concern with relationships between the individual and the social groups to which they belong. By adopting this perspective, sociologists have challenged arguments that human behaviour or beliefs are natural or inevitable and in so doing have provided a means by which disadvantaged groups have been able to question dominant power relationships.

There is no doubting the theoretical and political importance of sociology's emphasis on the causative influence of social relationships. When applied to environmental issues, this emphasis generates considerable insight into a host of questions ranging from why particular issues are considered important, while others are not, to how environmental change reflects institutional, political and economic arrangements. However, it is also an emphasis that is problematic in the sense that, at best, it can blind sociologists to the roles played by non-human processes and entities in environmental change and, at worst, treats human beings as if they exist somehow outside nature.

As stated in the introduction to this chapter, nature and the environment have largely been untheorised in sociology because their passive status and role have been taken for granted. This is not to say that they have been ignored altogether. The sociology of social movements may be considered the major sociological response to the surge of environmental activism in the 1960s and 1970s (Lockie 2004). An important feature of the sociology of social movements, however, was that the particular issues around which movements mobilised were seen as more or less irrelevant, and the causes of mobilisation were conceptualised in terms of exclusively social processes including macro-social structural change, contradictions within the capitalist mode of production, the inability of existing political institutions to adapt to change, conflict over access to resources, newly emerging political opportunities, and individual motives (Lockie 2004; see also Della Porta & Diani 1999). Rural sociologists, confronted with problems such as the impact of agricultural land degradation on farm livelihoods and productivity, have similarly been forced to confront environmental issues, but have done so by accepting scientific definitions of 'the problem' and getting on with the apparently more sociological task of working out how social factors – such as the imperatives of capitalist accumulation and the cultural milieu of farming practice – caused it (see Vanclay & Lawrence 1995). Rethinking our approach to environmental social theory in terms that do not rest on such a distinction between human society as the centre of agency and nature as the passive 'other' offers considerable opportunity to move beyond these approaches and engage more productively with contemporary biopolitical struggle (Goodman 1999).

The following section examines attempts to theorise nature and the environment from an explicitly sociological perspective. As it does so, it will look at the assumptions each attempt makes, either explicitly or implicitly, about: the fundamental nature of social life and the relationships between people and 'the environment'; how to develop valid 'knowledge' about these relationships; and the key features of social-environmental phenomena with which they are concerned and the research problems believed to be important.

KEY DEBATES

This section will be structured around four key questions that correspond to the main emphases of the variants of environmental social theory to be discussed here. The first concerns what we understand nature, the environment and environmental problems to be; the second, the causes of these environmental problems; the third, the role of science and technology in the production of nature, environmental problems and environmental knowledge; and the fourth, how we might go about addressing environmental problems and developing more 'ecologically rational' societies.

What is 'the environment'? Constructivism and materialism

Perhaps one of the most fundamental concepts within environmental sociology is the idea that 'the environment' is *socially constructed* (Hannigan 1995). This concept does not refer to the transformation of 'pristine' nature by humans into 'artificial' or 'built' environments, but to the ways in which our understandings of nature, the environment, and environmental problems are shaped by intrinsically social processes of knowledge generation and communication. This requires us to recognise that the terms we use to describe our environments (such as those used above) do not refer to universally applicable objective features of those environments but to socially valued categories and understandings that are liable to change across space, time and social groups. This can be illustrated by asking some simple questions. Why, for example, are 'wilderness values' so highly prized by Western environmental movements and seen to provide a scientific baseline on environmental condition by natural resource management agencies? Following from this, why is the state of ecosystems immediately before European colonisation in Australia, the United States and elsewhere in the 'New World' considered 'natural' despite millennia of ecological transformation at the hands of indigenous peoples? These values are not the reflection of absolute truths but of social conventions; that is, they are the result of a consensus between environmental groups, management agencies and scientists that the state of the pre-European ecosystem offers a convenient baseline against which to measure environmental degradation.

To the constructivist, there is no socially unmediated position from which to apprehend material reality. In its most extreme form (what is known as the 'strong program'), social constructivism argues that there is no reality whatsoever outside the symbolic world-building activities of humans and no way of knowing about that reality that is, in principle, any better or worse than any other way of knowing (Hannigan 1995). Science is simply more powerful than other ways of knowing. Rather than moving us ever and ever closer to 'the truth', science merely represents the perceived interests of dominant groups. Not surprisingly, the 'strong program' in social constructivism has been subjected to considerable criticism from the perspective of *materialism* (Mariyani-Squire 1999). From the materialist perspective, nature is attributed a pre-given objective reality that exists quite independently of human knowledge or action. Humans may learn about nature, and our interference may upset its balance to the detriment of particular species (including our own), but the basic material reality of nature is seen as something that cannot be changed. Those researchers taking a materialist perspective often accept 'scientific' statements about the state of environments and thence investigate why people damage them even though we 'know' the dangerous long-term consequences of this activity and, quite likely, have a number of solutions. Acting otherwise to the prescriptions of 'enlightened' scientists appears either irrational or impossible (Dunlap & Catton 1994).

The majority of social constructivists, however, accept the seemingly less extreme proposition that there *is* a pre-given and objective material reality that exists independently of humans, but that our knowledge of that reality is necessarily shaped by human categories, theories, projects, interests and power relationships (Greider & Garkovich 1994; Hannigan 1995). The relevance of this perspective is plainly evident in the example given above. Social constructivism does not suggest that imposing the category of 'wilderness' on landscapes that have been transformed by indigenous people provides a scientifically invalid baseline for measuring environmental change (although there may well be more useful baselines). What it does do is highlight the manner in which these categories deny any active role that may have been taken by indigenous people in transforming and managing those landscapes and the consequences this denial may have for contemporary struggles over cultural heritage, land rights, natural resource management, and so on (L'Oste-Brown et al. 2002). Social constructivism reminds us that the language we use to categorise and understand even the apparently 'objective' material reality of nature has political and cultural consequences for real people. We will return again to what this means for our understanding of science in later sections.

While the 'weak program' in social constructivism appears widely accepted, there is a degree of pragmatism in its response to the criticisms of materialism that might be considered unsatisfactory. The 'weak program' accepts the premises of materialism but offers no clear articulation of how the two perspectives may be accommodated. Instead, it simply focuses on those aspects of environmental issues amenable to semiotic analysis such as how issues are conceptualised by different groups, how the issues are represented in the mass media, and so on. The theoretical perspectives discussed in the following sections all attempt, in some way, to deal simultaneously with the insights of both the constructivist and materialist perspectives.

What causes 'environmental problems'? Capitalism and the production of nature

This section does not deal comprehensively with the possible causes of 'environmental problems' but, rather, introduces the theoretical perspective within environmental sociology – Ecological Marxism – that is most concerned with the identification of such causal relationships. The approach that Marxists have traditionally taken to dealing with a conceptual dilemma such as whether 'the environment' should be treated as a material reality or a social construct is to employ a method known as *dialectics*. Essentially, this means that instead of choosing one position in the debate between materialism and constructivism over the other, both are seen as characteristic of nature depending on the perspective from which it is viewed. In practical terms, taking a dialectical approach is not so different from the pragmatism of the 'weak program' in social constructivism. Both approaches acknowledge that 'the environment' is both a material reality and a symbolic construct. This has enabled Ecological Marxists, eco-feminists and others to analyse the ways in which the binary opposition of 'society' and 'nature' has been used to legitimate the subordination of groups – such as women and slaves – that have been defined as 'closer to nature' and thereby less human (Merchant 1987; Dickens 1996). However, thinking about the environment in dialectical terms points us towards a number of additional theoretical and empirical emphases.

Ecological Marxists, or political ecologists, use dialectical method to incorporate questions of agency (the capacity to make a difference) in human–environment relationships. Where classical Marxist theory adopts the modernist ideology of humanity liberating itself from 'brute nature' through the application of labour, Ecological Marxism argues that the lives of humans are conducted in relation to a nature that is both subject, and resistant, to transformation through human labour (Dickens 1996). Thus the dialectic is shifted from a concern with the relationship between the materiality of nature and its symbolic representation in human culture to a concern with the relationship between pre-existing 'laws' and potentialities of nature and the material production of nature through human labour. This leads Ecological Marxists to explain the alienation of humanity from nature through the capitalist labour process.

Capitalism, it is argued, provides powerful incentives for capitalist enterprises to discount or externalise social and environmental costs of production such as pollution (Schnaiberg & Gould 1994; O'Connor 1998). According to O'Connor, there are two basic contradictions within capitalist systems that ultimately undermine their economic and ecological sustainability:

1. *Demand crises:* the first 'contradiction of capitalism', as discussed in classical Marxist theory, is based on the tendency for individual enterprises to seek increased market share by increasing productivity and lowering costs. This they pursue by adopting technological innovations and shedding labour. However, when multiple enterprises do this the result is overproduction relative to consumer demand and declining prices.

2. *Cost crises:* the second 'contradiction of capitalism', derived from Ecological Marxism, is based on the need for individual enterprises to reduce costs so as to maintain competitiveness relative to other enterprises. Where social and environmental costs of production can be externalised, there is no incentive for individual enterprises to pay them, even though the long-term effect may be to undermine the resource base on which production depends and thus to raise average costs.

While capitalist enterprises are not bound by any inherent structural logic to externalise the environmental costs of production, they face the very real proposition that if they do not do so they will reduce their competitiveness and be replaced eventually by enterprises more prepared to do so. Strategies to avoid this situation include both state regulation and the development of markets for 'green' products, although both have been criticised by Ecological Marxists for adding a veneer of environmental responsibility to capitalist production while doing little to address the fundamental problems of resource overuse and pollution (Beder 1997).

Taking a dialectical approach and analysing problems from multiple perspectives makes good sense when confronted with difficult conceptual choices. But it is also an approach that risks, in relation to the conceptual dilemmas discussed here, paying lip service to the active and material role of nature while continuing to emphasise the causative role of classical macro-sociological abstractions such as 'society', 'capitalism' and 'the state' (Haraway 1991; FitzSimmons & Goodman 1998). In this sense, despite the value of their insights, many supposedly dialectical analyses may better be described as materialist in the simpler terms described in the previous section. Taking dialectical approaches to a more satisfactory conclusion requires incorporation of ideas such as the concept of a 'coevolutionary sociology' developed by Norgaard (1994) to examine the ways in which environments, knowledge, technology, values, and forms of organisation adapt to each other in unpredictable and potentially destructive ways. This will not be discussed in detail here. Instead, the chapter will move to consider a theoretical approach that dissolves altogether dichotomies between the social and natural, material and symbolic.

How do we know that 'environmental problems' exist? The sociology of scientific knowledge

As an enormously influential means for the generation of knowledge about 'the environment', science is an obvious target for sociological analysis. According to the 'strong program' in social constructivism, as stated above, science is no more objective than any other form of knowledge (e.g. indigenous or traditional knowledge), but it makes knowledge claims that reflect the interests and values of dominant groups. This sort of argument is regarded as deeply problematic by materialists, who counter that deconstructing the scientific basis on which knowledge of environmental problems is based potentially distracts our attention from what they argue are, in fact, 'real' and serious problems (e.g. Dunlap & Catton 1994). A compromise position might argue that while scientific work is as objective as possible, the priorities established for that research by governments and industry are likely to be those that accord with the interests of these groups. Yet there are good sociological and philosophical reasons to regard the claims of science to objectivity with some scepticism.

Numerous studies may now be found in the sociological literature of conflicting scientific findings in relation to environmental issues; of lay people being forced, in the face of scientific indifference, to conduct their own studies of environmental hazards; of scientifically developed environmental management strategies being proved ineffective; and so on. It is important, however, not simply to dismiss science on the basis of these examples but to consider, in more detail, why scientific observation is not a neutral but a theory-laden and political act. When choosing what to observe, scientists are influenced by what they believe already to be theoretically and socially relevant. When collecting data, they filter the infinite range of things that potentially could be observed through existing theory and experience. Patterns are observed in the data with which scientists are already familiar, and therefore scientific observation tends to support existing theory and existing solutions to social and environmental 'problems' (see Kuhn 1962; Feyerabend 1988).

Scientific knowledge is thus most accurately described not as invalid, but as partial and provisional. The question is, how then does scientific knowledge maintain its privileged position in the identification and management of environmental problems relative to other partial and provisional knowledges such as local and traditional knowledges? According to Latour (1987), the answer to this question lies in the ability of science to transform the world in its own image. What Latour means is that despite the generation of scientific knowledge through very localised research practices, that knowledge is recorded and communicated in a manner that facilitates its application outside the laboratory. Represented as universal, or generalisable, scientific knowledge is used to exert influence on people and landscapes, often at considerable spatial and temporal remove from the research sites at which that knowledge was generated, and in the process defining, monitoring and modifying the key attributes of those people and landscapes. This demonstrates that the distinction between the social construction and the material reality of 'the environment' is in fact a false one. At the same time that scientific knowledge of nature is embedded in human systems of language and symbols, that knowledge is developed through relationships with the organisms, substances and patterns of nature that contribute to the transformation of both science and the phenomena under investigation in the process.

Further to this, the sociology of scientific knowledge demonstrates that the expression of human agency, or influence, is enabled by technologies that capture, preserve and transfer knowledge and materials through space and time. The networks of the social are thus not exclusively human, but hybrids of people, nature and technologies, all of which may, intentionally or otherwise, resist their enrolment. In an era of so-called 'globalisation', such a conclusion appears almost commonsense. At the same time that the spatial and temporal extension of social relationships characteristic of globalisation is enabled only by developments in transport, telecommunications and computing (Latour 1999), the environmental implications of global capitalism are also becoming increasingly clear (Beck 1992). But this is also a conclusion that appears scandalous in light of sociology's traditional refusal to countenance a causal role for anything other than 'social facts' in the explanation of social phenomena.

The sociology of scientific knowledge is often criticised for trying to apply concepts such as power and agency – traditionally applied by sociologists only to people and social groups – to non-human nature and technologies. Hacking (1999), for example, argues that clear distinctions must be drawn between humans and non-humans on the basis of human capacity for language, consciousness, reflection and intentional resistance. This is not contested by the sociology of scientific knowledge. Instead, agency and power are conceptualised not as properties of individuals (human or otherwise) but as the outcomes of interactions within a network. Conceptualising power and agency as properties of relationships and not of the individuals involved is, in fact, a widely accepted proposition among social theorists (e.g. Giddens 1984; Foucault 1986; Hindess 1996). For Foucault (1980, 1986), power is unstable, reversible, pervasive and, as often as not, accompanied by resistance and evasion. It follows that power may take many forms, at times concentrated and hierarchical and at times dispersed (Hindess 1996). Sociologists of scientific knowledge thus argue that rather than attempting to define power and agency, or to whom or what they should be attributed, in theoretical terms, they should be treated as empirical research questions (Callon & Law 1995; Latour 1999).

The theoretical challenge of the sociology of scientific knowledge to mainstream sociology may be seen to lie not so much in its injunction to consider the roles played by non-human nature and technologies in the networks of the social, but in the injunction shared with other branches of sociological theory to conceptualise key sociological concepts in more relational terms (Latour 2000; Murdoch 2001). In this respect, the sociology of scientific knowledge has a great deal to offer mainstream sociology in terms of the analytical tools and concepts it has developed with which to conduct research into the material manifestations of power and agency. These have been illustrated in this section through discussion of the universalisation of scientific knowledge (for more detail see Latour 1993; Law 1994; or for more introductory material Murdoch 1995, 1997).

The main weakness in the sociology of scientific knowledge for the purposes of this chapter is the lack of guidance it might offer in terms of practical interventions in natural resource management. While this perspective is very well placed to analyse the ways in which power and agency are enabled and constrained by social environmental networks, and always presupposes that things could be otherwise, it has very little to say regarding how social networks *ought* to be arranged (Latour 2000). For this reason, the final perspective outlined in this chapter is one that deals explicitly with such questions.

What do we do about 'environmental problems'? Deliberation and democracy

Deliberative theory is concerned with the integration of empirical, moral and critical aspects of research into mechanisms for making collective choices (Dryzek 1996). As such, it provides a rationale for the trend in natural resource management over the last two decades to adopt a discourse of citizen participation and democratisation, and guidance as to how the social sciences may support these processes. Deliberative theory certainly does not go as far as the sociology of scientific knowledge in its theoretical challenge to dichotomies between the social and the natural, the symbolic and the material, but it does focus in quite a practical way on the relationships between the construction of environmental knowledge and material intervention in environmental management.

Deliberative theory is concerned with the promotion of communicatively rational deliberation; that is, with negotiation over social and environmental issues that is oriented towards the attainment of consensus through free and unconstrained debate among communicatively competent equals, and thus founded solely on the merits of arguments rather than on the defence of particular interests or points of view (Habermas 1984; Dryzek 1990). The role of deliberation is not to establish universal standards of 'right' and 'wrong', or to find the one 'correct' answer to a dispute or problem, but to arrive at decisions that participants believe fair and reasonable (Miller 1992). Decisions may be based as much on participants' assessments of appropriate procedures or norms as on their assessment of empirical 'facts'. Communicatively rational deliberation is based on the assumption that while it is not possible in many circumstances for all claims to be satisfied (e.g. in use of a resource), it is possible for people's views 'to be swayed by rational arguments and to lay aside particular interests and opinions in deference to overall fairness and the common interest of the collectivity' (Miller 1992: 56).

Communicatively rational deliberation is not, of course, the principal collective choice mechanism in Western societies, which are dominated instead by instrumental rationality. Instrumental, or means–end, rationality promotes the manipulation of objects for short-term human gain, the discounting of future benefits and of intrinsic value in nature, and a reductionist outlook that ignores complexity and interdependence. By contrast, according to Dryzek (1987), communicatively rational deliberation has real potential to promote more 'ecologically rational' decision-making by encouraging decision-makers to consider options holistically and look beyond self-interest to collective and ecological well-being, by decentralising decision-making and making social choice processes more sensitive to ecological feedback signals, and by involving more participants in what appear to be supra-local issues (Dryzek 1987, 1992).

An obvious question to ask here is what makes deliberative theory anything more than wishful thinking? The answer, according to Dryzek (1996), lies in empirical examples of the differences between situations where natural resources have been managed sustainably over long periods and those that have not. In relation to sustainable resource management situations, he argues, it is evident that those involved have developed ways to communicate and interact with each other. They also have developed the ability to learn about whom to trust and the effects that their actions have on each other and on the resource; and they have developed norms, patterns of reciprocity, institutions, and so on to solve problems and conflicts. Individuals are not constituted in such cases as either rational individualists or rational ecologists, but as members of a web of relationships from which identity and coordination are drawn. Situations characterised by instrumental rationality, by contrast, are subject to the same imperatives towards resource degradation discussed earlier in this chapter in relation to capitalism.

The social sciences may contribute to communicatively rational deliberation in a number of ways. The principles of free and unconstrained debate may be used to evaluate and critique existing collective choice mechanisms and to suggest alternatives. In this respect, the social sciences can go beyond advocating wider citizen participation as an end in itself and help to uncover the way in which instrumental rationality is embedded in institutions and decision-making processes in ways that are taken for granted but which yet undermine communicative rationality. The social sciences can also be used to clarify the impacts of proposed courses of action on different social groups, and to document and represent the interests and aspirations of groups affected by those proposals. The aim here is not to adopt the objectifying gaze of the 'expert' social scientist but to use the methods of the social sciences to support deliberation over the goals, impacts and management of proposed change.

Despite the clear identification of roles for the social sciences in the promotion of communicative rationality, deliberative theory offers little by way of specific methods or conceptual tools with which to do this. Despite some major theoretical differences, it can be seen that the ability of the sociology of scientific knowledge to examine the ways in which particular rationalities (instrumental, communicative or ecological) are represented as universal, and hence made to appear inevitable through their inscription in technologies and ways of doing things, offers opportunities to add considerable empirical depth to this analysis.

FUTURE DIRECTIONS

Reflecting its title – social nature – this chapter has argued that resolution of the key theoretical debates in environmental sociology requires abandonment of dichotomies between nature and society, the material and the symbolic, as embedded within mainstream social theory (see also Braun & Castree 1998). Further, following the most overt attempts to do this leads us to a reformulation of basic sociological concepts such as power and agency in terms that see these conceptualised as the outcome of relationships within networks rather than as the properties of individuals. By shifting the focus of sociological analysis from a search for *the* centre of power to exploration of the multiple ways in which power and agency may be expressed and contested, it becomes possible to consider more seriously the active roles played by nature and technology in the networks of the social.

Reviewing a number of key theoretical perspectives in environmental sociology has, however, also highlighted considerable scope for dialogue and integration across these perspectives. While it is possible to suggest that communicatively rational deliberation offers an empirically and theoretically grounded social and political goal to guide the application of concepts from the sociology of scientific knowledge, Ecological Marxism and social constructivism, there remain a number of differences between these perspectives that should not be overlooked.

It is also important to remember here that the chapter has not been comprehensive in its treatment of legitimate theoretical perspectives in environmental sociology. While many of those perspectives that have not been addressed align with the conceptual debates that have dominated this chapter, others raise additional questions and opportunities. Eco-feminism, in particular, highlights dichotomies between male and female that have also been implicated in the project of modernity. Certainly, it is possible to suggest that one opportunity for integration lies in applying the approach taken in the sociology of scientific knowledge to treating such dimensions of difference as empirical research questions rather than as theoretically fixed categories. But it is also important to remember the eco-feminist argument that as differences are materialised in the embodied experience of real people, including researchers, they contribute to the necessarily partial, or 'situated', knowledge that researchers bring to the research process. This suggests that some potentially very fruitful future directions for environmental social theory might lie in more extensive dialogue between non-dichotomous theoretical approaches and more consideration, in particular, of the methodological implications of these approaches.

Finally, it has been suggested that social theory that deals seriously with the organisms, substances and patterns of nature offers opportunities to engage with contemporary biopolitics movements in productive new ways. However, beyond the promotion of communicative rationality little has yet been written on how this might happen. Articulating how more sophisticated environmental social theory might contribute to constructive engagement with opposition social movements and mainstream natural resource management agencies alike suggests itself as a worthwhile priority.

Discussion Questions

1. What is the challenge presented to mainstream sociology by environmental sociology?
2. In what ways might the environment be described as a social construct?
3. What environmental issues are in the mass media at the present time? How are they represented? What are the consequences of these representations for the groups concerned?

4. What are the two contradictions underlying capitalism and what are their environmental implications?

5. How might sociological understanding of capitalist systems contribute to environmental sustainability?

6. Can science offer an objective basis on which to make environmental decisions? In what ways is science influenced by social and institutional factors?

7. What revisions are suggested by the sociology of scientific knowledge to sociological concepts of power and agency? Are these changes convincing?

8. What possibilities exist for sociology to intervene, in a practical way, in the ways in which we construct and manage environments?

Glossary of Terms

Capitalism: a mode of production characterised by private production of goods and services, using wage labour, for sale as commodities.

Communicative rationality: consensus reached through free and unconstrained debate among communicatively competent equals, and thus founded solely on the merits of arguments rather than on the defence of particular interests or points of view.

Deliberative democracy: collective choice mechanisms based on communicatively rational deliberation.

Dialectical: apparently opposing pairs of concepts that may be resolved at a higher level of analysis.

Ecological Marxism: a branch of Marxist theory concerned with the analysis of environmental questions from a Marxist perspective.

Ecological rationality: environmental decision-making based on communicative rationality.

Materialism: a theoretical perspective based on the proposition that nature and social relationships have an objective reality that exists independently of human knowledge or beliefs.

Social constructivism/constructionism: a theoretical perspective concerned with the ways in which human knowledge and behaviour are shaped through social processes including systems of ideas and meanings.

Sociology of scientific knowledge: (also known as Actor Network Theory) a sociological perspective based on ethnographic studies of scientific research. This perspective advocates the treatment of power and agency as the outcomes of interactions within social networks and, therefore, making no *a priori* assumptions about the relative influence of humans and non-humans within those networks.

Theory: a coherent group of general propositions used to explain a class of phenomena. Main schools of thought within a discipline.

References

Beck, U. 1992, *Risk Society: Towards a new modernity*, London: Sage.

Beder, S. 1997, *Global Spin: The corporate assault on environmentalism*, Melbourne: Scribe Publications.

Braun, B., and N. Castree (eds) 1998, *Remaking Reality: Nature at the millennium*, London: Routledge.

Callon, M., and J. Law 1995, 'Agency and the hybrid collectif', *South Atlantic Quarterly* 94(2): 481–507.

Della Porta, D., and M. Diani 1999, *Social Movements: An introduction*, Oxford: Blackwell.

Dickens, P. 1996, *Reconstructing Nature: Alienation, emancipation, and the division of labour*, London: Routledge.

Dryzek, J. 1987, *Rational Ecology: Environment and political economy*, Oxford: Basil Blackwell.

Dryzek, J. 1990, *Discursive Democracy: Politics, policy and political science*, Cambridge University Press.

Dryzek, J. 1992, 'Ecology and discursive democracy: beyond liberal capitalism and the administrative state', *Capitalism, Nature, Socialism* 3(2): 18–42.

Dryzek, J. 1996, 'Foundations of environmental political economy: the search for *Homo Ecologicus?*', *New Political Economy* 1(1): 27–40.

Dunlap, R., and W. Catton 1994, 'Struggling with human exemptionalism: the rise, decline and revitalization of environmental sociology', *American Sociologist* 25(1): 5–30.

Durkheim, E. 1966, *The Rules of Sociological Method*, 8th edn, New York: Free Press.

Feyerabend, P. 1988, *Against Method*, rev. edn, London: Verso.

FitzSimmons, M., and D. Goodman 1998, 'Incorporating nature: environmental narratives and the reproduction of food'. In Braun and Castree, *Remaking Reality*.

Foucault, M. 1980, *Power/Knowledge*, Brighton, UK: Harvester.

Foucault, M. 1986, 'Disciplinary power and subjection'. In S. Lukes (ed.) *Power*, Oxford: Blackwell.

Giddens, A. 1984, *The Constitution of Society: Outline of the theory of structuration*, Berkeley CA: University of California Press.

Goodman, D. 1999, 'Agro-Food studies in the "age of ecology": nature, corporeality, bio-politics', *Sociologia Ruralis* 39(1): 17–38.

Greider, T., and L. Garkovich 1994, 'Landscapes: the social construction of nature and the environment', *Rural Sociology* 59(1): 1–24.

Habermas, J. 1984, *The Theory of Communicative Action*, vol. 1, Cambridge: Polity Press.

Hacking, I. 1999, *The Social Construction of What?* London: Harvard University Press.

Hannigan, J. 1995, *Environmental Sociology: A social constructionist perspective*, London: Routledge.

Haraway, D. 1991, *Simians, Cyborgs and Women: The reinvention of nature*, New York: Routledge.

Hindess, B. 1996, *Discourses of Power: From Hobbes to Foucault*, Oxford: Blackwell.

Jary, D., and J. Jary 1991, *Dictionary of Sociology*, Glasgow: Harper Collins.

Kuhn, T. 1962, *The Structure of Scientific Revolutions*, University of Chicago Press.

Latour, B. 1987, *Science in Action: How to follow scientists and engineers through society*. Cambridge MA: Harvard University Press.

Latour, B. 1993, *We Have Never Been Modern*, Cambridge MA: Harvard University Press.

Latour, B. 1999, 'On recalling ANT'. In J. Law and J. Hassard (eds) *Actor Network Theory and After*, Oxford: Blackwell.

Latour, B. 2000, 'When things strike back: a possible contribution of "science studies" to the social sciences', *British Journal of Sociology* 51(1): 107–23.

Law, J. 1994, *Organizing Modernity*, Oxford: Blackwell.

Lockie, S. 2004, in press, 'Collective agency, non-human causality and environmental social movements: a case study of the Australian "landcare movement"', *Journal of Sociology*.

L'Oste-Brown, S., L. Godwin and C. Porter in association with the Bowen Basin Aboriginal Steering Committee 2002, *Towards an Indigenous Social and Cultural Landscape of the Bowen Basin*, Brisbane: Queensland Department of Environment.

Mariyani-Squire, E. 1999, 'Social constructivism: a flawed debate over conceptual foundations', *Capitalism, Nature, Socialism* 10: 97–125.

Merchant, C. 1987, *The Death of Nature: Women, ecology and the scientific revolution*, San Francisco CA: Harper & Row.

Miller, D. 1992, 'Deliberative democracy and social choice', *Political Studies* 40: 56–67.

Mills, C.W. 1959, *The Sociological Imagination*, New York: Oxford University Press.

Murdoch, J. 1995, 'Actor-networks and the evolution of economic forms: combining description and explanation in theories of regulation, flexible specialization, and networks', *Environment and Planning A* 27: 731–58.

Murdoch, J. 1997, 'Inhuman/nonhuman/human: actor-network theory and the prospects for a nondualistic and symmetrical perspective on nature and society', *Environment and Planning D: Society and Space* 15: 731–56.

Murdoch, J. 2001, 'Environmental sociology and the ecological challenge: some insights from actor network theory', *Sociology* 35(1): 111–33.

Norgaard, R. 1994, *Development Betrayed: The end of progress and a coevolutionary revisioning of the future*, London: Routledge.

O'Connor, J. 1998, *Natural Causes: Essays in Ecological Marxism*, New York: Guilford Press.

Pepper, D. 1984, *The Roots of Modern Environmentalism*, London: Routledge.

Schnaiberg, A., and K. Gould 1994, *Environment and Society: The enduring conflict*, New York: St Martin's Press.

Turner, B. 1996, *The Body and Society: Explorations in social theory*, 2nd edn, London: Sage.

Vanclay, F., and G. Lawrence 1995, *The Environmental Imperative: Eco-social concerns for Australian agriculture*, Rockhampton, Qld: CQU Press.

CHAPTER THREE

GENDER, ECO-FEMINISM AND THE ENVIRONMENT

Val Plumwood

Eco-feminism has several major aims. One project is to connect feminist and ecological perspectives, thought and movements, developing 'a feminism that is ecological and an ecology that is feminist', in Ynestra King's (1989) words. Eco-feminist thinking grew from criticism of sexism in the green movement and lack of ecological consciousness in the women's movement into a critique opposing all forms of oppression. Eco-feminist thinkers have also explored conceptual and cultural connections between women and nature, and applied feminist power analyses to problems in environmental philosophy.

Much feminist and eco-feminist philosophical critique has focused on mind/body dualism and the denial of embodiment as the key background for the environmental failure of Western culture. Early eco-feminism challenged dualised conceptions of spirit as transcendent male deity and developed alternative eco-feminist philosophies and spiritualities of embodiment and immanence (Ruether 1975; King 1981; Spretnak 1982). Contemporary ecological feminism often draws on an analysis of mind/body, spirit/matter and male/female dualisms or deep conceptual splits to understand the contribution of gender to the forms of culture and economic rationality that bring contemporary societies into ecological danger zones – the ecological crisis.

BACKGROUND TO THE ISSUES

Eco-feminist historical scholars Rosemary Ruether (1975) and Carolyn Merchant (1980) and others established major conceptual connections between women and nature in Western culture. Women in this culture have been historically associated with the supposedly 'lower' order of nature, with animality, materiality and physicality, and men with the contrasting 'higher' order of mind, reason and culture. As Karen Warren

(2000) puts it, nature has been feminised and women naturalised, so that understanding these connections is necessary to understanding their respective oppressions. Critical eco-feminism sees culture/nature (or human/nature) dualism as the key to the ecological failings of Western culture.

These different orders of nature and culture are constructed in terms of a dualism of oppositional spheres, which are arranged as emphatically distanced (or hyper-separated), and contrasted as higher and lower, as centre to periphery, as active to passive, as mind to mindlessness, and as subject to object of knowledge (Stepan 1993). The 'higher' side of this dualistic division, associated with mind, spirit and reason, is linked to elite male humans who have transcended physical labour, while women (and slaves) are associated with the body, animality, emotionality and deficiency in reason. Human/nature dualism is a key, linking part of the network of culture/nature, spirit/matter, mind/body and reason/nature dualisms that have shaped Western culture, and is an active force in contemporary life. If the basic structure of thought the dominant culture has inherited is based on highly gendered, linked pairs of dualistic contrasts – of male/female, mind/body, reason/emotion, and humanity/animality – rethinking of all these categories is required.

Eco-feminist critique of Western culture and the historical development of human apartness

For critical eco-feminism, human/nature dualism is a Western-based cultural formation going back thousands of years that sees the essentially human as part of a radically separate order of reason, mind, or consciousness, set apart from the lower order that comprises the body, the animal and the pre-human (Lloyd 1984; Spelman 1988; Plumwood 1993). Supposedly inferior orders of humanity, such as women, slaves and ethnic Others ('barbarians'), partake of this lower sphere to a greater degree, through their supposedly lesser participation in reason and greater participation in lower 'animal' elements such as embodiment and emotionality. Human/nature dualism conceives the human as not only superior to but as different in kind from the non-human, which as a lower sphere exists as a mere resource for the higher human one. This ideology has been functional for Western culture in enabling it to exploit the non-human world and so-called 'primitive' cultures with less constraint, but it also creates dangerous illusions in denying embeddedness in and dependency on nature, which we see in our denial of human inclusion in the food web and in our poor response to the ecological crisis.

Ancient Greek society was androcentric, anthropocentric and built on slavery. Its democracy privileged the 30 per cent of the population allowed to vote – the elite males who saw themselves as representing spirit, mind and reason. The remainder, women and slaves, were identified with material, bodily labour and led highly separated, confined and devalued lives. Correspondingly, in ancient Greek philosophy, the earthly, material world of materiality and embodiment is seen as not only inferior but corrupting, and those who leave it behind at death pass to a higher and purer realm of immateriality beyond the earth. Plato's philosophy is 'dematerialising', treating reason – lodged in a pure realm of immaterial, timeless ideas – as opposed to or threatened by the corrupted material world of 'coming to be and passing away', the biological world of nature and the body (Lloyd 1984; Spelman 1988; Plumwood 1993). The hyper-separation and devaluation of the body and matter are not confined to ancient philosophy but are widespread in Western culture and were inherited by the dominant Western religious movements of Christianity. In the spirit of the classical Greek tradition of earth denial, Christian ideals of salvation and transcendence of the material subordinated the 'unimportant' earthly world of nature and material life to the next world of heaven, the immaterial celestial world beyond the earth, where non-humans could never go. In the ascent to a better world of spirit beyond earthly, embodied life, matter would ultimately be conquered by the opposing elements of spirit and reason in a process of dematerialisation.

With modernity, reason as modern science began to rival and replace religion as the dominant belief system. Western science replaced but also built on this earlier religious foundation, transforming the idea of conquering nature as death by subordinating nature to the realm of scientific law and technology. Modern science, now with religious status, has tended to inherit and update rather than supersede these oppositional and supremacist ideals of rationality and humanity. In the scientific fantasy of mastery, the new human task becomes that of remoulding nature to conform to the dictates of reason to achieve salvation – here on earth rather than in heaven – as freedom from death and bodily limitation. This project of controlling and rationalising nature has involved both the technological-industrial conquest of nature made possible by reductionist science and the geographical conquests of empire.

The idea of human apartness emphasised in culture, religion and science was of course shockingly challenged by Charles Darwin in his argument on the descent of species that humans evolved from non-human species. But these insights of continuity and kinship with other life forms (the real scandal of Darwin's thought) remain only superficially absorbed in the dominant culture, even by scientists. The traditional scientific project of technological control is justified by continuing to think of humans as a special superior species, set apart and entitled to manipulate and commodify the earth for their own benefit. Against the evidence that animals like birds are just as evolved as humans, human-centred culture assumes that humans are the apex of creation, more intelligent, more communicative, more important and much more evolved than other species (Midgley 1983).

This sense of apartness is challenged by the new science of ecology that stresses the importance of biosphere services and ecological processes, and the dependence of humans on a healthy biosphere and a thriving more-than-human world. But the influence of dematerialising philosophers like Plato and Descartes who treat consciousness, rather than embodiment, as the basis of human identity, continues in a false consciousness and mode of life which fails to situate human identity, human life and human places in material and ecological terms. Thus dematerialising, globalising culture understands 'our place', in accord with mind/body dualism, as our conscious homeplace, nation or identity group rather than in terms of the many global places we draw upon in the global market for the material support for our lives, especially the places of our ecological footprint. The global fishing industry kills hundreds of thousands of penguins, dolphins and other cetaceans yearly in nets used in remote places, yet few fish-eaters are aware of this slaughter or press for alternatives that do not degrade the environment. Global markets are dematerialising because they prioritise abstract relationships and in them we lose touch with the real material conditions that support our lives. Will we be able to change this false consciousness in time to survive? This question hangs in the balance at the present time.

Varieties of eco-feminism

Eco-feminism is diverse and draws on divergent political ideals. Some authors, such as Rosemary Ruether, extend communitarian socialism through attention to the oppression of women and the non-human. Mary Daly and some other radical feminists develop their eco-feminism as part of a gynocentric worldview which emphasises gender oppression, masculinity and patriarchal culture as the privileged factor in explaining cultural failings like environmental degradation (Daly 1978; Collard 1989). These basic kinds of divisions persist in later eco-feminisms. Other varieties, such as critical eco-feminism, emerge oppositionally in relation to debates discussed below.

For social eco-feminisms the key category for understanding is not that of production associated with men's labour, privileged in Marxist frameworks, but rather that of reproduction, associated with women's labour in the home and various support activities outside. Reproduction can be understood as including both women's labour and the non-human ecological conditions on which production depends (Merchant 1980). This crucial enabling category is neglected and denied in the Western tradition and its economic and cultural theories and practices. The distorted view of the world enshrined in dominant theories corresponds to the perspective of a dominant gender, class and species category, that of elite men, foregrounding their own contribution as 'production' and backgrounding what others, human and non-human, do to make that possible. The agency of the world is theirs; they are the ones that make things happen, others merely the 'hired hands', wives, secretaries, support staff, which includes, on a very lowly rung, the biosphere. Nature and women are passive.

Consequent illusions and blindspots about agency and about how processes of (material) production depend on reproduction are crystallised in capitalism as a form of economic life that simultaneously assumes and disavows ecology, the body and reproduction. Capitalism is seen as a distorted form of reason that maximises the throughput of nature (in production) to generate wealth and profit without allowing for its renewal (or reproduction). These 'ecologically irrational' forms of reason, which assume but simultaneously deny their ecological base, are strongly implicated in the ecological crisis (Mies & Shiva 1993; Plumwood 2002).

Anti-colonial and ecosocial eco-feminisms are combined in analyses of Western capitalism as the conquest of society and nature by a colonising and enclosing form of development or reason (Mies & Shiva 1993; Plumwood 1993). Colonisation involves both conquest of other races and nations and conquest of associated parts of nature, both considered inferior and open to improvement via invasion and domination because identified as 'closer to nature', as an earlier, more primitive stage of life (Mies & Shiva 1993; Plumwood 1993). The dual connection of women and other subordinated groups with nature and of male elites with reason is the key to the fundamental colonising problematic of Western culture. Mies & Shiva blend anti-colonial and anti-capitalist positions by analysing capitalism as imposing the hegemony of a Western development model based on the master subject, White Man, which is eurocentric, androcentric, and anthropocentric (human-centred). Anti-colonial forms of eco-feminism provide a detailed account of how environmental crises and non-human forms of oppression are linked to human oppression and injustice. The distortion of rationality through the dualistic constructions associated with domination is the foundation of the distorted Western concept of development and progress, or 'maldevelopment', as Vandana Shiva (1994) calls it.

KEY DEBATES

The essentialism debate within feminism and eco-feminism

How can we now affirm women's efforts in the sphere of reproduction, as nurturers and mediators of nature, as eco-feminism now seems to require? Liberal or Artemisian feminism, gynocentric feminism and critical forms of eco-feminism part company on how to resolve this dilemma, giving rise to the essentialism and 'earth mother' debates discussed below.

These questions are subsidiary to the larger question within feminism: how should women now approach the ideals of the feminine and the contributions of past women? When the Greek goddess Artemis sought affirmation from her father Zeus by asking for all her brother Apollo's stuff, especially his weapons, she did nothing to assert her own identity and value as a woman. This strategy fails to challenge the model, for its male-centred or its destructive features, and lacks solidarity because it aims to escape the normal fate of woman, rather than to improve it. The idea that gender liberation means women joining men in an unproblematised sphere of culture (equated with humanity) seen as opposed to nature is characteristic of what is often called 'liberal' feminism, really a conservative, Artemisian form of feminism that aims at equality within an unquestioned androcentric ideal or model.

Many feminists have urged that women need to do more than Artemis for real equality, because the androcentric ideals of humanity and culture Artemisians would join subtly presuppose and privilege maleness and its associated ideals and characteristics, real or assumed. A few women may succeed, but most women will not do well or achieve equality under androcentric regimes – as in contemporary public life and in the workplace. And, eco-feminists add, neither women nor men will do well in a regime which subtly denies the ecological basis of human life. Most feminists and eco-feminists agree that the dominant models of culture and humanity must be challenged, both for women and for nature.

Gynocentric feminists particularly want to celebrate aspects of women's lives (Daly 1978). Other feminists have identified several problems, under the heading 'essentialism', for the affirmation of women's identity and value. If affirmation covers women's traditional role, there is the problem of the inseparability of much of this role from subordination, the fine line between peacefulness and passivity or obedience. An uncritical celebration of women's nurturance or passivity is not compatible with equality (Spelman 1988; Plumwood 1993). And many women have not conformed to such ideals, which are tied to specific cultures, classes and conditions. It is doubtfully possible to find any general characteristics that hold good of all women, especially women of different cultures, and the assumption of a universal women's nature ('essentialism') is either bio-deterministic or, more usually, hegemonic, promoting a dominant model and ignoring subordinated cultures and practices (Carlassare 1994; Spelman 1988). Both these problems count against the idea of simply celebrating women's role, as nurturers for example.

A resolution of the debate that avoids essentialism need not exclude all affirmation. A critical affirmation which attends to cultural context and power is required in the context of subordination and cultural variation. But against the Artemisians, it can be said that affirmation is important, because to be effective politically women need to be able to express solidarity with other women (Carlassare 1994; Sturgeon 1997), and to correct the illusions of agency by restoring some sense of the value of their contribution – and, eco-feminists would add, that of nature also. The critique of 'essentialism' rightly cautions against uncritical affirmation and cultural universalism, but should not rule out critical affirmation and solidarity.

How should women now relate to nature?

How can modern women who wish to be liberated now relate to the sphere in which they were traditionally included, the sphere of nature? Is the connection to nature just an insulting old idea that women should now throw off completely to join men in a culture supposedly liberated from nature?

The project of all varieties of feminism is the recognition of women's equal humanity, but there are many different analyses of basic concepts and ideals of the human and of what equality involves. The eco-feminist critique of the gendered dualisms of Western culture outlined above suggests several requirements and problems for rethinking women's inclusion in humanity. Both ecological and eco-feminist analysis can be seen as indirectly challenging human (ecological) and gender hyper-separation. Many eco-feminists, like other feminists, find objectionable women's traditional treatment as less than fully human and their consequent inclusion in the separate and inferior sphere of nature as opposed to culture. But there are two distinct ways to go about challenging this construction, corresponding to gynocentric and critical forms of eco-feminism. One way would retain the traditional gender separation and the idea that women are part of nature, but reverse the traditional ordering to proclaim that nature is superior to culture, or glorify women because of their closer relationship to nature, especially through their position as mothers and nurturers. It is not men but women who are to present the new ideals of humanity.

This gynocentric reversal strategy has drawn much criticism from feminists, especially Artemisians, who see it as continuing women's imprisonment in the sphere of reproductivity and family and their exclusion from the true humanity of the public world of culture, work and public life. Gynocentrism challenges the traditional presumed inferiority of the sphere of nature and women, but it does not challenge the idea that women (but not men) are part of it. Nor does it challenge women's exclusion from culture, or the way these spheres are conceived as hyper-separated and highly exclusive of each other. But substitution of a traditional female for a traditional male model of the human does not address the issue of distortion of both male and female roles and models by dualistic gender constructions.

A more thorough challenge, a critical eco-feminism (Plumwood 1993), argues that women are no more 'part of nature' or 'closer to nature' than men are – both men and women reside in both nature and culture. We must rethink hyper-separations, both the opposition and polarisation of men and women and that of humanity and nature. Like gynocentric eco-feminism, critical eco-feminism disputes the inferiority of the sphere of nature, but unlike gynocentric eco-feminism denies its exclusive link to women.

The resulting program is both feminist and eco-feminist; with feminism, a critical eco-feminism rejects women's exclusion from culture, defined dualistically as the province of elite men who are seen as above the base material sphere of nature and daily life and able to transcend it (in creativity or production) through their greater share in reason or agency (enterprise). The supposed creative transcendence of culture is built on the denial and subjection of those assigned to the sphere of reproduction or nature – both women and the non-human – whose supporting role is invisible but essential. The dominant culture's distancing from and backgrounding of nature leads to a sense of independence from it that is dangerous and illusory.

Also like social and anti-colonial feminism, critical eco-feminisms reject the exclusion and distancing of the 'ideal' human type, the 'productive' male, from the sphere of nature, ecology and reproduction to which women have been confined, and the idea that this hyper-separated concept of culture is the true mark and home of the human. Both men and women must participate in this sphere, in household, childcare and earthcare, as well as in the sphere previously designated as male, that of culture (Plumwood 1993, 2002; Merchant 2003). Like ecology, eco-feminism promotes an ecological consciousness which insists that a truly human life is embedded in both nature and culture, which are not hyper-separated spheres as Artemisian feminism assumes. Critical eco-feminisms challenge nature/culture and human/nature dualisms, and ask both feminism and philosophy to rethink the concepts of both *woman and the human* in ecological terms that are respectful of non-human difference, sensitive to human continuity with non-human nature, and attentive to the embodiment of all life and the embedment of human culture in the more than human world.

Ecological Animalism versus Ontological Veganism

There is an intense and unresolved debate within environmental philosophy between positions extending ethical concern only to (some) animals and positions emphasising ethical concern for ecological systems and all living things. This choice between animals or ecology is reflected in eco-feminism in a debate over hunting and vegetarianism, as a complex debate between gynocentric versus critical eco-feminist positions. Both sides of this debate oppose contemporary animal abuse and express solidarity with and care for animals as part of disrupting human/nature dualism and apartness. Critical eco-feminism shows how a third position, an Ecological Animalism, can draw the animal and ecology positions closer together, and also resist the gender-stereotyping and essentialist appeal of some feminisms to a gentle, passive and peaceful 'woman's nature' that is culturally invariant and incapable of hunting or killing animals.

Feminist thinking, critical eco-feminists urge (Sturgeon 1997; Plumwood 2002), must recognise the full diversity of women's lives in diverse cultures where the meaning of hunting and eating other animals is variable. In contemporary US and modern urban contexts, recreational hunting of wild animals often has a strongly gendered meaning of establishing masculinity as toughness, aggression, male bonding based on excluding women, and suppressing 'soft' emotions such as sympathy – all features mobilised in warfare. But hunting can have a very different meaning in the context of foraging societies where animals are respected but some animal food is essential for survival in a limited ecosystem. If the meaning of hunting is culturally variable, the dangers of gynocentric and culturally invariant concepts of women reside especially in their potential for fostering cultural hegemony via falsely universalising the experience and perspectives of privileged (e.g. white urban) women and obscuring those of other women.

For example, from an anti-colonial eco-feminist perspective, the sweeping assumption that 'women' do not hunt and that female-led 'gathering' societies were vegan or plant-based (Collard 1989; Adams 1994: 107) presupposes a culturally universalist model of human life as modern Western urban life, along with a gendered dualism of foraging activities in which the mixed forms of hunting and gathering encountered in many indigenous societies (e.g. Australian Aboriginal societies where women routinely hunt smaller animals) are denied and disappeared. The gynocentric feminist tendency to universalise and decontextualise 'woman' and to privilege explanations blaming men and masculinity means that the cost-cutting economic rationality behind contemporary animal debasement is neglected in Adams' theoretical framework.

Both gynocentric and critical eco-feminisms agree that modern factory farming abuses animals. Farm animals have never been as cruelly confined or slaughtered in such numbers in all human history as they are under modern 'rational' farming systems. The factory farm today is crowded and stinking, full of suffering animals that are sprayed with pesticides and fattened on diets of growth hormones, antibiotics and drugs. Two hundred and fifty thousand laying hens may be confined within a single building with hardly space to move or stretch, unable to peck or nest. But people opposing these abuses face an important set of choices about how to theorise and extrapolate their opposition. In particular, they have to choose whether to opt for theories of animal ethics and ontology that emphasise discontinuity and set human life apart from animals and ecology (Adams 1993, 1994), or theories that emphasise human continuity with other life forms and situate both human and animal life within an ethically and ecologically conceived universe. They also have to decide between theories that attribute the abuse of animals in farming systems to individual weaknesses such as greed for meat, or instead to capitalist systems of rationality and profitability that minimise animals' share of the world and reduce them to commodities or resources for consumption.

This choice appears in two eco-feminist theories that challenge, in quite different ways, the ideology of human/animal dualism. Ontological Veganism is a theory that advocates individual consumer abstention from all use of animals as the only real alternative to factory farming and the leading means of defending animals against its wrongs. But another theory which also supports animal defence, Ecological Animalism, more thoroughly disrupts human/nature dualism, and is better than Ontological Veganism for environmental awareness, for human liberation, and for animal activism itself.

Ecological Animalism supports and celebrates animals and encourages a dialogical ethics of sharing the world and negotiation or partnership between humans and animals, while undertaking a re-evaluation of human identity that affirms inclusion in animal and ecological spheres. Ecological Animalism is a context-sensitive semi-vegetarian position, which advocates great reductions in first-world meat-eating and opposes reductive and disrespectful conceptions and treatments of animals, especially as seen in factory farming.

As we have seen, the dominant position that is deeply entrenched in Western culture constructs a great gulf or dualism between humans on one side and animals and nature generally on the other. Human/nature dualism conceives the human essence as mind or spirit, not body; humans are inside culture but 'outside nature' – not conceived ecologically as part of a system of exchange of nutrition and never available as food, for example, to other animals. Non-humans are seen in polarised and reductive terms as outside ethics and culture, and as mere bodies, reducible to food (Midgley 1983; Luke 1995; Weston 1996). Ecological Animalism aims to disrupt this deep historical dualism by resituating humans in ecological terms at the same time as it resituates non-humans in ethical and cultural terms. It affirms an ecological universe of mutual use, and sees humans and animals as mutually available for respectful use in conditions of equality. Ecological Animalism insists we must consider context to express care for both animals and ecology, and to acknowledge at the same time different cultures and individuals in different ecological contexts, differing nutritional situations and needs.

Ontological Veganism has numerous problems for both theory and activism, animal equality and ecology. It ties strategy, philosophy and personal commitment tightly to personal veganism, abstention from eating and using animals as a form of individual action (Adams 1993, 1994, 2003). It insists that neither humans nor animals should ever be conceived as edible or even as usable, confirming the treatment of humans as 'outside nature' that is part of human/nature dualism, and blocking any reconception of animals and humans in fully ecological terms. Because it is indiscriminate in proscribing all forms of animal use as having the same (low) moral status, it fails to provide philosophical guidance for animal activism that would prioritise action on factory farming over less abusive forms of farming. Its universalism makes it highly ethnocentric, universalising a privileged 'consumer' perspective, ignoring contexts other than contemporary Western urban ones, or treating them as minor, deviant 'exceptions' to or departures from what it takes to be the ideal or norm (Adams 1993, 1994; Eaton 2002). Although it claims to oppose the dominant ideology of apartness, it remains subtly human-centred because it does not fully challenge human/nature dualism, but rather attempts to extend human status and privilege to a bigger class of 'semi-humans' who, like humans themselves, are conceived as above the non-conscious sphere and 'outside nature', beyond ecology and beyond use, especially use in the food chain. In doing so it stays within the system of human/nature dualism and denial that prevents the dominant culture from recognising its ecological embeddedness and places it increasingly at ecological risk.

Ontological Veganism's subtle endorsement of human/nature dualism and discontinuity also emerges in its treatment of predation and its account of the nature/culture relationship. Predation is often demonised as an instrumental practice bringing unnecessary pain and suffering to an otherwise peaceful vegan world of female gathering. But if instrumentalism is not the same as simply making use of something, and even less thinking of making use of it (ontologising it as edible), predation is not necessarily a reductive or instrumental practice, especially if it finds effective ways to recognise that the other is more than 'meat'. Ecologically, Ontological Vegans present predation as an unfortunate exception and animals, like women, as always victims: fewer than 20 per cent of animals, Adams tells us, are predators (Adams 1993, 2003) – a claim that again draws on a strong discontinuity between plants and animals. In this way it is suggested that predation is unnatural and fundamentally eliminable. But percentage tallies of carnivorous species are no guide to the importance of predation in an ecosystem or its potential eliminability.

An Ecological Animalist would say that it is not predation as such that is the problem but what certain social systems make of predation. They would agree that hunting is a harmful, unnecessary and highly gendered practice within some social contexts, but reject any general demonisation of hunting or predation, which would raise serious problems about indigenous cultures and about flow-on from humans to animals. Any attempt to condemn predation in general ontological terms will inevitably rub off onto predatory animals (including both carnivorous and omnivorous animals), and any attempt to separate predation completely from human identity will also serve to reinforce once again

the Western tradition's hyper-separation of our nature from that of animals, and its treatment of indigenous cultures as animal-like. This is
another paradox, since it is one of the aims of the vegan theory to affirm
our kinship and solidarity with animals, but here its demonisation of
predation has the opposite effect of implying that the world would be
a better place without predatory animals. Ontological Vegans hope to
avoid this paradox, but their attempts to do so reveal clearly that their
worldview rests on a dualistic account of human identity.

The main move Ontological Vegans make to minimise the significance of predation and block the problematic transfer of their anti-
predation stance from humans to animals is to argue that human
predation is situated in culture while animal predation is situated in
nature (Adams 1993: 206; Moriarty & Woods 1997). One paradox is
that animal activists who have stressed our continuity with and similar-
ity with animals in order to ground our obligation to extend ethics to
them now stress their complete dissimilarity and membership of a separate order, as inhabitants of nature not culture, in order to avoid a
flow-on to animals of demonising all predation. Embracing the claim
that humans 'don't live in nature' in order to block the disquieting and
problem-creating parallel between human hunting and animal predation introduces a cure which is worse than the disease and which is
basically incompatible with any form of ecological consciousness. The
claim that humans are not a part of natural ecosystems is on a collision course with the most fundamental point of ecological understanding because it denies the fundamental ecological insight that human
culture is embedded in ecological systems and dependent on nature. It
also denies an important insight many students of animals have rightly
stressed: that culture, learning and choice are not unique to the human
and that non-human animals also have culture. In fact Woods and
Moriarty's solution rests on a thoroughly dualistic and hyper-separated
understanding of human identity and of the terms 'nature' and 'culture'.
In order to attain the desired human/animal separation, nature must be
'pure' nature, 'strictly biological', and culture conceived as 'pure' culture, no longer in or of nature: an activity is no longer natural if it
shows any cultural influence, and culture is completely disembedded
from nature, 'held aloft on a cloud in the air'.

Of course Ontological Vegans are right to object to any simple naturalisation of human hunting and meat-eating. For Ecological Animalism, both the claim that meat-eating is in nature rather than culture and the counter-claim that it is in culture and therefore not in
nature are wrong and are the product of indefensible hyper-separated

ways of conceptualising both these categories that are characteristic of human/nature dualism. It is only if we employ these hyper-separated senses that the distinction between nature and culture can be used to block the flow-on problem that condemning predation in sweeping terms also condemns animal predation. For critical eco-feminism, any form of human eating (and many forms of non-human eating) is situated in both nature and culture – in nature as a biologically necessary determinable and in a specific culture as a determinate form subject to individual and social choice and practice. Both naturalising and culturalising conceptual schemes are inadequate to deal with the problem, since both sides deny the way our lives weave together and crisscross narratives of culture and nature, and the way our food choices are shaped and constrained both by our social and by our ecological context.

FUTURE DIRECTIONS

Conventional animalist and conventional ecological theories as they have evolved in the last four decades have each challenged only one side of this double dualist dynamic, and they have each challenged different sides, with the result that they have developed in highly conflictual and incompatible ways. Although each project has a kind of egalitarianism between the human and non-human in mind, their partial analyses place them on a collision course. *The ecology movement has been situating humans as animals, embodied inside ecological systems of mutual use, of food and energy exchange, just as the animal defence movement has been trying to expand an extension to animals of the (dualistic) human privilege of being conceived as outside these systems.* Many vegans seem to believe that ecology can be ignored and that the food web is an invention of hamburger companies, while the ecological side often retains the human-centred resource view of animals and scientistic resistance to seeing animals as individuals with life stories of attachment, struggle and tragedy not unlike our own, refusing to apply ethical thinking to the non-human sphere. A more double-sided understanding of and challenge to human/nature dualism can help us move on towards a synthesis, a more integrated and less conflictual theory of animals and ecology, if not yet a unified one.

Eco-feminists are increasingly taking on the environmental theory generated by traditional male philosophers to sketch a different vision of human relationship to the non-human world, for example in the intense and long-running debate among environmental theorists over human-centredness (anthropocentrism). Environmental philosophers have debated whether our relationships with nature are distorted, destructive and irrational because our framework of thought is human-centred, or whether human-centredness is, as some urge, the only possible way for humans to relate to the world. This stereotypical debate presupposes a false choice between human and non-human interests and sides of the environmental problem (Plumwood 1998). This false choice between self and other, human and non-human, that is so entrenched in the framework of environmental ethics discussion is well exemplified in work by historian of environmental thought Peter Hay (2002). Hay begins by identifying the environmental problematic completely with what is really only part of it, compassion for other life forms. In his first chapter introducing the 'ecological impulse' and motivating the environment movement, Hay proceeds immediately to identify the paradigm of environmental activism as wilderness defence. 'The cornerstone of the environment movement', he writes, 'may well be the impulse to defend . . . the existential interests of other life-forms' (Hay 2002: 25).

Eco-feminism reformulates both human and non-human sides of the problem as an outcome or expression of the *human/nature dualism* that in Western culture deforms and hyper-separates *both sides* of what it splits apart (Plumwood 2002; Merchant 2003). This analysis escapes the false choices between human and non-human, instrumental and intrinsic, prudential and ethical, self and other, because it sees our failures in situating non-humans ethically and our prudential failures in failing to situate our own lives ecologically as closely and interactively linked. Countering the human/nature dualism associated with human-centredness gives us two tasks: (re)situating humans in ecological terms and non-humans in ethical terms. The first is apparently the more urgent and self-evident, the task of prudence or care for self, while the other is presented as optional, the inessential sphere of ethics or care for the other. But this is an error; the two tasks are interconnected, and cannot be addressed properly in isolation from each other.

To the extent that we hyper-separate ourselves from nature and reduce it conceptually in order to justify domination, we not only lose the ability to empathise and to see the non-human sphere in ethical terms, but we also get a false sense of our own character and location that includes an illusory sense of agency and autonomy. The reductive mindset that refuses to see the non-human other in the richer terms appropriate to ethics (Weston 1996, 2004; Plumwood 2002) licenses supposedly 'purely instrumental' relationships that distort our perceptions and enframings, impoverish our relations, and make us insensitive to limits, dependencies and interconnections – which are thus in turn a prudential hazard to self. The eco-feminist focus on the larger political and historical context of human/nature dualism can give us a fuller, more integrated and coherent conception of the environmental problematic, broadening the narrow 'deep' focus on non-human and wilderness issues to represent more closely the full range of issues and concerns in real environmental struggles.

Discussion Questions

1. What are the major concerns of eco-feminism?
2. Why and how have women been seen as closer to nature than men?
3. Are contemporary women now liberated from nature?
4. Outline some differences among eco-feminists. Why and to what does critical eco-feminism present a more thorough challenge?
5. Why do eco-feminists think the conceptual framework of Western culture is implicated in the environmental crisis?
6. Are women less capable of hunting than men? Can this question be answered without reference to a social context?
7. What is the real problem with essentialism and how is it displayed in the eco-feminist debate on veganism?
8. Is eating situated in nature or in culture? How do personal context, social context and ecological context bear on how we can and should eat?
9. What is human/nature dualism and why does ecological animalism present a more thorough challenge to it?

Glossary of Terms

Androcentrism (of culture): male-centred culture which sees women as having a radically different and inferior nature to men, privileges masculinity in social structures, and treats women as deviant or exceptional.

Anthropocentrism: human-centred culture, treating non-humans as a radically distanced, inferior and reduced category which is passive and available to service humans; treatment of other species as highly separate and inferior in a way analogous to eurocentric treatment of other races as highly separate and inferior and androcentric treatment of women as highly separate and inferior.

Artemisian feminism: another term for 'liberal feminism', based on the story of the goddess Artemis, which 'has tended to accept the basic structures of existing political and economic institutions, pressing hardest on the need to make them accessible to women' (Nicholson 1986: 24). Also termed the feminism of uncritical equality (Plumwood 1993).

Cultural universalism: the extension of the norms and assumptions of one (dominant) culture to all cultures.

Dualism: the creation of two radically separated or polarised orders based on hyper-separation between a dominant and a subordinated class or group.

Essentialism: the implication that women's nature is universal and culturally invariant, usually because it is seen as biologically determined.

Gynocentrism: a position which aims to reverse the dominant order, taking women to be at the cultural centre in the same way as men are in androcentrism. It sees woman's nature as universal and positive, and gender domination as the only or most significant axis of oppression, to which other kinds are to be reduced.

Human/nature dualism: the idea that humans belong to a superior order set apart from the inferior order of animality and the non-human.

Hyper-separation: an emphatic form of separation, distinct from simple separation or non-identity, which stresses the differences between two polarised groups and denies continuity and overlap.

Instrumentalism: values and modes of relationship within which something is valued only for use as a means to a further end, rather than in itself or for its own sake.

Naturalisation (of oppression, inferiority):   organising concepts so that inferior treatment (of an oppressed group) seems natural and inevitable; relatedly, the inclusion of women in the category of nature.

Transcendence: rising above and beyond a lower order.

References

Adams, C.J. 1990, *The Sexual Politics of Meat: A feminist-vegetarian critical theory*, New York: Continuum.
Adams, C. 1993, 'The feminist traffic in animals'. In G. Gaard (ed.) *Ecofeminism*, Philadelphia PA: Temple University Press.
Adams, C.J. 1994, *Neither Man nor Beast: Feminism and the Defense of Animals*, New York: Continuum.
Adams, C. 2003, *The Pornography of Meat*, New York: Continuum.
Carlassare, E. 1994, 'Essentialism in ecofeminist discourse'. In C. Merchant (ed.) *Ecology: Key concepts in critical theory*, Atlantic Highlands NJ: Humanities Press.
Collard, A. with J. Contrucci 1989, *Rape of the Wild: Man's violence against animals and the earth*, Bloomington IN: Indiana University Press.
Daly, M. 1978, *Gyn/Ecology: The metaethics of radical feminism*, London: Women's Press.
Eaton, D. 2002, 'Incorporating the other: Val Plumwood's integration of ethical frameworks', *Ethics and the Environment* 7(2): 153–80.
Hay, P. 2002, *Main Currents of Environmental Thought*, Sydney: UNSW University Press.
King, Y. 1981, 'Feminism and Revolt', *Heresies* 4(1): 12–26.
Lloyd, G. 1984, *The Man of Reason*, London: Methuen.

Luke, B. 1995, 'Solidarity across diversity: a pluralistic rapprochement of envi-
    ronmentalism and animal liberation', *Social Theory and Practice* 21(2);
    also in R.S. Gottlieb 1997 (ed.) *The Ecological Community*, London:
    Routledge, pp. 333–58.
Mellor, M. 1997, *Feminism and Ecology*, Cambridge: Polity Press.
Merchant, C. 1980, *The Death of Nature*, London: Wildwood House.
Merchant, C. 2003, *Reinventing Eden: The fate of nature in Western culture*, New
    York: Routledge.
Midgley, M. 1983, *Animals and Why They Matter*, Athens GA: University of
    Georgia Press.
Mies, M., and V. Shiva 1993, *Ecofeminism*, London: Zed Books.
Moriarty, P.V., and M. Woods 1997, 'Hunting≠Predation', *Environmental
    Ethics* 19 (Winter): 391–404.
Nicholson, L. 1986, *Gender and History: The limits of social theory in the age of
    the family*, New York: Columbia University Press.
Plumwood, V. 1993, *Feminism and the Mastery of Nature*, London: Routledge.
Plumwood, V. 1998, 'Paths beyond human-centredness: lessons from liberation
    struggles'. In A. Weston (ed.) *An Invitation to Environmental Philosophy*,
    Oxford/New York: Oxford University Press, pp. 69–106.
Plumwood, V. 2000, 'Integrating ethical frameworks for animals, humans and
    nature: a critical feminist eco-socialist analysis', *Ethics and the Environment*
    5(3): 1–38.
Plumwood, V. 2002, *Environmental Culture: The ecological crisis of reason*,
    London: Routledge.
Ruether, R. 1975, *New Woman, New Earth*, Minneapolis MN: Seabury.
Shiva, V. 1994, 'The Seed and the Earth'. In V. Shiva and M. Mies (eds)
    Close to Home: Women reconnect ecology, health and development, London:
    Earthscan.
Spelman, E. 1988, *The Inessential Woman*, Boston MA: Beacon.
Spretnak, C. 1982, *The Politics of Women's Spirituality*, New York: Doubleday.
Stepan, N.L. 1993, 'Race and gender: the role of analogy in science'. In
    S. Harding (ed.) *The Racial Economy of Science*, Indianapolis IN: Indiana
    University Press, pp. 359–76.
Sturgeon, N. 1997, *Ecofeminist Natures: Race, gender, feminist theory and political
    action*, New York: Routledge.
Warren, K.J. 1990, 'The power and promise of ecological feminism', *Environ-
    mental Ethics* 12(2): 121–46.
Warren, K.J. 2000, *Ecofeminist Philosophy: A Western perspective on what it is and
    why it matters*, New York: Rowman & Littlefield.
Weston, A. 1996, 'Self-validating reduction: toward a theory of environmental
    devaluation', *Environmental Ethics* 18: 115–32.
Weston, A. 2004, 'Multi-centrism: a manifesto', *Environmental Ethics* 26(1):
    25–40.

CHAPTER FOUR

ANIMALS, 'NATURE' AND HUMAN INTERESTS

Lyle Munro

The emergence of environmental sociology in recent decades has legitimated the study of nature–society relations within the discipline of sociology. However, within environmental sociology itself, the nature/society couplet has been defined in quite narrow terms, excluding for the most part our relations with non-human animals.

The key debates in this chapter are suggested in the title 'Animals, "Nature" and Human Interests', where each of the three concepts are problematised in various ways. Animal rights supporters argue that there are two kinds of animals – human and non-human – and that both have rights and interests as sentient beings; they believe, however, that the dominant ideology of 'speciesism' enables humans to exploit non-human animals as commodities to be eaten, displayed, hunted and dissected for their benefit. Environmentalists believe that nature (as in the natural world) needs to be distinguished from 'nature' or the *Umwelt* (a German word which literally means the world around us); this 'nature' is the constructed nature of gardens, zoos, aquariums, natural history museums and so on. And finally, the term 'human interests' signals the possibility of conflicts of interest between different humans (e.g. environmentalists and factory farmers, vivisectionists and animal liberationists, hunters and 'hunt saboteurs') as well as the notion of human interests versus the interests of non-human animals (e.g. our desire to see megafauna in zoos versus the captive animal's desire to run wild, or our taste for meat versus the suffering of individual animals in slaughterhouses or in the live animal export trade).

This chapter examines some of the controversies associated with these issues from the perspectives of different interests. It concludes with a discussion of the ethical and social issues associated with genetically engineering animals which might feature as future animals in food, research, entertainment and numerous other contexts in which humans use other animals.

BACKGROUND TO THE ISSUES

Eleven years ago, the *Journal of Social Issues* ran a special edition on 'The role of animals in human society' in which the issue editor, a psychologist (Plous 1993), remarked on how the topic of animals and their 'rights' was not seen by many of his colleagues in the psychological fraternity as an appropriate one for the journal. A typical comment was to ask whether the journal would be covering the 'rights of strawberries' next. A similar response to the issue of animal rights can be traced back to 1792 when Mary Wollstonecraft's *A Vindication of the Rights of Women* was satirised by Thomas Taylor in an essay which replaced 'women' with 'beasts' in order to imply – if women are meant to have rights, why not beasts or for that matter strawberries?

The role of animals in society has been debated for centuries, mainly by philosophers. In more recent times, academic disciplines from anthropology to zoology have contributed different perspectives on animal–human relations. Plous, for example, identifies five contexts in which animals play important roles in human societies: companionship, economics, the environment, health, and morality. He goes on to list some less sanitised uses of animals by humans:

> Animal products are found not only in the foods we eat and the clothes we wear, but in the walls of our homes (eg in sheetrock and wallpaper adhesive), in kitchen and bathroom floors (eg in linoleum, ceramic tiles, and floor wax), in toiletries (eg in soap, perfume, deodorants and cosmetics), in the streets of our cities (eg in asphalt binders) and the cars we drive (eg in brake fluid, upholstery, and car wax). Animal products are also used in a variety of paints, plastics, textiles, and machinery oils. (Plous 1993: 2)

Industries which profit from these products, and hundreds like them, want to ensure that people know as little as possible about their origins. Thus we are presented with sanitised products such as the pork chop, the lamb cutlet, the hamburger – that is, animal parts – which are unlikely to remind us of the whole animal. Similarly, slaughterhouses, animal laboratories and so on are discreetly hidden away out of sight or behind closed doors so as not to upset consumers of animal products, or worse, invite attacks from animal rights activists.

During the last two centuries, there has been a growing concern about our treatment of animals that has culminated in the formation of animal welfare organisations like the RSPCA as well as more radical animal rights groups such as People for the Ethical Treatment of Animals (PETA) in the United States. Groups such as these seek to make visible via publicity and direct action campaigns the plight of animals on farms, in labs, in the wild, in zoos and so on. One observer of animal–human relations, the historian E.S. Turner (1992), is impressed that so many creatures now have their own pressure groups. On the other hand, there are so many contexts in which animals suffer that it is difficult for people who care about animals to know where to start. Nonetheless, there is a broad consensus in the animal movement on what constitutes the worst examples of 'speciesism' – a social problem not unlike sexism, racism and rankism.

Sociological surveys of supporters of the animal rights movement in Australia and the United States reveal a number of (ab)uses of animals which respondents believe are morally reprehensible. These are shown in Table 4.1 below. What is most interesting about these results is that there is strong agreement among animal rights supporters on what constitutes the worst forms of speciesism from a range of practices where humans use animals or their products. There is a strong consensus in these countries that the worst practices involve animal experimentation, hunting and farming practices, while meat-eating (ranked 9) and keeping animals in zoos (ranked 10) are seen as less objectionable. This chapter therefore focuses on these five issues (1–10 in Table 4.1) as the most controversial in the contested field of animal rights politics. The chapter describes these issues from the perspective of individuals, groups and social movements involved in contesting or defending our relations with non-human animals in various contexts including the recent controversy over the genetic engineering of animals. The animal rights movement identifies all of these practices as constituting animal abuse and constructs speciesism in these contexts as a social problem on a par with other social problems such as the abuse of children, women, and racial and ethnic minorities.

The rest of the chapter explains why the animal rights movement challenges speciesism in its seminal campaigns against vivisection, factory farming and bloodsports – campaigns which nonetheless are contested by virulent counter-movements as well as by 'God-fearing carnivores' in the related issue of vegetarianism.

Many people see environmentalists, animal activists and vegetarians as kindred spirits; however, these groups are characterised by important ideological differences in the way they view lab animals, farm animals and especially wild animals. In the case of the latter, the issues of recreational hunting and the role of zoos – for example in relation to endangered species – highlight the controversies and dilemmas for

Table 4.1 *Mean scores of Australian and American activists' attitudes towards the use of animals*

		Mean rating[a]	
		ANZFAS Australia	Richards USA
1	Using steel-jawed leg-hold traps to capture wild animals	1.02	1.06
2	Using animals in cosmetic and beauty product experiments	1.05	1.13
3	Killing an animal to make a fur coat	1.16	1.17
4	Selling unclaimed dogs from animal shelters for use in medical experiments	1.19	1.29
5	Hunting wild animals with guns	1.32	1.49
6	Exposing an animal to a disease as part of a medical experiment	1.33	1.62
7	Raising cattle for food in feedlots	1.34	1.75
8	Using horses for jump/steeple racing	1.79	2.68
9	Eating meat	2.81	2.74
10	Keeping animals in zoos	3.08	3.02
11	Raising cattle for food on open range or pastures	3.48	3.31
12	Killing rats in residential area	4.93	4.24
13	Killing cockroaches in a residential area	5.35	5.34
14	Keeping a dog or cat as a pet	6.31	6.49
15	De-sexing a pet	6.64	6.62
		Mean 2.85 (SD2.03)	Mean 2.93 (SD1.92)

Note: [a] rating scale values range from 1 (extremely wrong) to 7 (not at all wrong).
Source: Table in Munro (2001: 133)

vested interests on both sides of these debates. While philosophers have been the most prominent commentators on the animal rights debate, in recent years sociological concepts and theories have contributed to the analysis of human–animal relations in journals such as *Society and Animals*.

KEY DEBATES

The environmental controversies discussed below focus on the fauna side of nature and include several constructions of non-human animals: research animals, food animals, trophy/display animals, captive animals and genetically engineered animals. The section begins with a general debate about the status of animals and their treatment in sociological discourse.

The speciesist–sentient-centric continuum

The sociologists Alan Wolfe (1993) and Luke Martell (1994) represent both ends of the speciesist–sentient-centric continuum. Wolfe's book *The Human Difference: Animals, Computers and the Necessity of Social Science* is a defence of traditional sociology's indifference to nature. He argues that sociology was born between two worlds, the supernatural (dominated by religious ideas) and the animal (dominated by natural-istic ideas). 'Sociology was a product of the notion that humans were, and ought to be, at the centre of our attention. Its founding thinkers agreed that the line between humans and the worlds surrounding them was dangerously thin and that, therefore, humans and their accomplish-ments required a special defense' (Wolfe 1993: 3–4). Wolfe argues that sociology is therefore necessarily anthropocentric since it is concerned with what makes us different from the animate (animals and nature) and the inanimate (computers and artificial intelligence). Wolfe iden-tifies three radical attempts at 'putting nature first': animal rights, deep ecology, and Gaia.

In Wolfe's critique, Gaia is more hostile to human impacts on nature than deep ecologists, who in turn manage to make mainstream ani-mal liberationists seem quite moderate. Yet Wolfe is highly critical of animal rights philosophy, which he claims would result in a world without *fantasy, excitement and creativity* (1993: 87). He argues that the extension of rights to non-human beings, who possess neither agency nor understanding, devalues the concept of human rights, a point often made by lay persons. While many would agree with his strict anthropocentric/speciesist stance, most environmentalists and all ani-mal liberationists would want to strenuously challenge it.

Wolfe's argument seems to be that we have to be cruel to ani-mals if we are to enjoy the good life: experiencing fantasy (seeing megafauna in zoos and circuses); excitement (such as in sport hunting); and creativity (quintessentially in the genetic engineering of animals). Martell's (1994) position as a sentient-centric sociologist is diametri-cally opposed to Wolfe's speciesist stance. We look at some of the main contexts where humans (ab)use animals in the rest of the chapter.

Research animals: animal experimentation versus vivisection

Many people believe that animal experimentation or vivisection is 'the most sensitive and polarizing issue in the animal rights controversy' (Jasper & Nelkin 1992: 137). The controversy thus far has been char-acterised by the rhetoric of hostility and vilification over the issue of whether or not animals should be used in research. The debate is one of the longest-running controversies in the animal movement's history; the most recent arguments from the protagonists appeared in *Scientific American* (Mukerjee 1997), which devoted its cover story to a full exploration of the issue. Similarly, Plous (1993) outlined the controversy among psychologists in the *Journal of Social Issues*. On a broader level, the controversy often appears in ethics and philosophy anthologies in an attempt to encourage debate and understanding in relation to human versus animal interests and rights. One recent scholarly work on the subject is *Brute Science*, a book by LaFollette & Shanks (1996), which examines the moral and practical issues from both sides of the issue. The authors summarise the arguments as follows:

The case for animal experimentation

- most medical advances have resulted, directly or indirectly, from research using animals;
- there would be serious consequences for human health and well-being if the research stopped;
- cell and tissue cultures, and computer simulations are not satisfactory alternatives to animal experimentation;
- animal experimentation is scientifically justified because human and non-human animals are biologically similar (LaFollette & Shanks 1996: 10).

The case against animal experimentation

- the role of animal research in reducing mortality rates has been exaggerated;
- the contribution of animal experimentation to interventionist medicine has been exaggerated;
- the results of research on animals have sometimes been very misleading;
- the practice of using animals in research is morally repugnant (LaFollette & Shanks 1996: 16).

Sociologists Arluke & Groves (1998) have analysed the culture of animal research, while Groves (1997) has written about the role of emotion in the controversy over lab animals. Both are perceptive ethnographies of the work of animal researchers and their critics. Groves, for example, notes that in this highly charged issue, animal rights people rationalise their emotionality, while animal researchers emotionalise their rationality (1997: 14).

Food animals: intensive farming versus factory farming; meat versus mercy

What farmers happily call intensive farming, animal liberationists denounce as 'factory farming'; what carnivores call meat, animal rights/vegetarian activists call animal flesh. Many environmentalists are opposed to agribusiness and the industrialisation of animal food production because large-scale animal industries like piggeries damage the environment; most 'animal lovers' find factory farms morally objectionable for reasons of cruelty, while vegetarians have a range of reasons why they avoid meat including health, gustatory, and ecological concerns as well as compassion for animals. There is therefore a broad consensus or common cause which environmental and animal rights/vegetarian activists agree on in relation to factory farming. Nonetheless, there is debate on how the relationship between environmentalists and animal activists can be described – optimistically as kindred spirits or more pessimistically as unreliable friends? (see Hargrove 1992). Environmentalists typically focus their efforts on saving wildlife species rather than individual domesticated animals, who are the concern of animal liberationists. Historically, at least in Australia, both have worked together in campaigns to save the koala, whose survival was threatened in the 19th century by the trade in koala furs. The early preservationists found the animal protectors' talk of 'bush babies' embarrassingly sentimental, but as MacCulloch (1993) points out, it was their construction of the pathos of cruelty to animals which appealed to the public and literally saved the koalas' skins.

On the issue of meat-eating, strict animal rights/vegetarian activists and environmentalists do not see eye to eye. Furthermore, within the animal movement itself, there is much disagreement about whether meat constitutes what the vegetarian/animal rights advocate Carol Adams (1990: 70) claims is 'the most oppressive and institutionalized violence against animals'. The results of the survey in Table 4.1 above suggest otherwise; here meat-eating is ranked 9th – well down on the hierarchy of speciesism – in the list of fifteen practices which animal defenders object to. Even so, an increasing number of writers have pointed out the health and environmental costs of large-scale food animal production as opposed to the eco-friendly impact of vegetarianism (e.g. Fiddes 1991; Rifkin 1992; Penman 1996).

The meaning of meat became a topic of widespread public interest during the mid-1990s when two 'vegetarian/animal rights activists from hell' challenged the fast food giant McDonald's and its practices

during the McLibel Trial in London (see Vidal 1997). A similar controversy occurred with the so-called 'veggie libel laws' in the United States
when the Texas Cattlemen's Association launched a multimillion dollar lawsuit against the television personality Oprah Winfrey when she
allegedly made disparaging comments about hamburgers. Films like
Babe and *Chicken Run* have also caused agribusiness interests to worry
about the sale of their animal products to a public (temporarily) turned
off meat-eating by the plight of animals portrayed in such films.

Given the relatively small number of vegetarians in the world, it
is perhaps surprising that the growth of a vegetarian, animal-friendly
sensibility should be viewed with alarm by the meat industry. However, according to the philosopher R.G. Frey (1983), there are at least
twenty-one reasons why we should be concerned with the prospect of
a mass conversion to vegetarianism; these range from the demise of
the huge meat industry and the economies it supports to the effects
on the restaurant trade. He concludes that the animal welfare benefits would pale into insignificance compared to the costs of a meat-free
world, though vegetarian-animal rights advocates claim that these costs
would be short-lived. The reality is that there seems little prospect of an
increase in vegetarianism large enough to threaten the dominant meat-
eating culture; however, the 'civilising process' (Elias 1978) suggests
that cruelty and violence towards animals may be declining with the
possibility that vegetarianism may become the long-term dietary norm
for humans in the future. Like other issues discussed in this chapter,
attempting to predict future outcomes is a hazardous enterprise. Frey's
position may be the most likely in that he believes the 'concerned individual' of the future will only eat meat produced under humane conditions such as traditional farming, and eschew meat produced by morally
objectionable practices.

Trophy animals: recreational hunting or bloodsport?

Many environmentalists, unlike strict animal rightists and vegetarians,
have no qualms about either hunting wild animals or eating their meat.
The sociologist-hunter Jan Dizard (1994), in the aptly titled *Going
Wild: Hunting, Animal Rights and the Contested Meaning of Nature*, can-
vasses the conflicting views of animal rights activists, environmentalists
and hunters in a dispute over wildlife management. In this particular
case during the early 1990s, the managers introduced a nine-day deer
hunt to cull the deer who were eating out the new growth that supported Boston's main water supply. Dizard argued that the deer-hunters
had a better understanding of nature than the animal defenders, who

he claimed were essentially afflicted by the 'Bambi syndrome', a label that is sympathetically deconstructed by a strong critic of sport hunting, Matt Cartmill (1993). Like Elias, Cartmill suggests that the human–animal boundary is gradually becoming blurred and as a result, certain kinds of hunters are an endangered species. These 'endangered' hunters would not include traditional, subsistence hunters, but rather recreational hunters and so-called trophy hunters.

It is in England, the home of the original 'animal lovers' who founded the RSPCA in 1824 and organisations like the League Against Cruel Sports in 1924, that the hunting controversy is at its most intense. The issue concerns the rights of recreational hunters on horseback to use dogs to hunt foxes in the English countryside versus the animal protectors and 'hunt saboteurs' who claim the sport is cruel and should be banned. Fox-hunters make a number of claims to justify their sport: it's an ecologically friendly means of culling wild animals; it provides communities with employment opportunities; and most importantly, it's a traditional English countryside pastime which urban-based 'bunny-huggers' should not be allowed to destroy. Hunters recently joined in a mass protest in 1997 to defend their way of life against their critics in 'the cat and dog brigade' who they claim, as Dizard did in the Boston deer-hunt controversy, know little about nature beyond their pet-keeping. Nonetheless, public opinion in England and elsewhere appears to be against recreational hunting or 'bloodsports', as the animal movement calls it.

Trophy hunting and its extreme version in 'canned hunting' are bloodsports which even respectable hunters deplore. People who simply shoot animals for 'fun' or make a trophy of the body parts of animals are not seen as legitimate hunters unless they actually hunt, rather than shoot the animal; the most extreme form of shooting animals for fun is the practice of 'canned hunting' whereby the shooter, for a price, is provided with a wild animal – perhaps a bear or an ex-zoo elephant – to shoot in captivity without the need to hunt, stalk and kill the animal. This is possibly the most extreme form of the commercial exploitation of animals for trivial purposes discussed so far. Yet according to Wolfe (1993), the practice could be justified because it satisfies human cultural needs such as fantasy, excitement and even creativity, in so far as 'canned hunting' is an innovative addition to the world of bloodsports.

Display animals: zoo animals as captive animals

While zoos are not the most controversial of issues for animal rights activists, there is nonetheless a lively debate between advocates and critics of zoological gardens, as they used to be called. Zoological gardens were originally built in the large cities of Europe and elsewhere as places of spectacle and entertainment. They evolved from the menageries of antiquity which were kept by royalty. After the French Revolution in 1789, zoos were built in many capital cities to promote civic pride, education and entertainment. According to a recent history of zoos in 19th-century America, the modern zoo is evolving into a Bio Park which 'will combine elements of existing zoos, aquariums, natural history museums, botanical gardens, arboretums, and ethnological and anthropological museums to create a holistic form of bioexhibitry' (Robinson 1996: xi). The author claims that the Bio Park puts an end to the 'unnatural separation' of flora and fauna because the concept is about exhibiting plants and animals, not plants *or* animals. Yet no one can deny that the Bio Park is constructed 'nature', not nature as it exists in the wild.

There is now a large literature on the pros and cons of zoos written by scholars in disciplines ranging from anthropology to zoology. Among the most recent pro-zoo books are Bostock (1993) and de Courcy (1995), who argue that modern zoos fulfil four broad functions: entertainment, research, education and conservation. Anti-zoo books, for example by Mullin & Marvin (1987) and Malamud (1998), challenge these functions with counter-arguments: entertainment for humans should not be at the expense of animals who have to be caged; research on captive animals cannot be as valid as observations of wild animals in the natural state; learning about captive animals is really a form of mis-education as the animals are perceived as enslaved; the prospect of saving endangered species with captive breeding programs is at best marginal. Both sides of the debate are represented in Norton and colleagues' *Ethics of the Ark: Zoos, animal welfare and wildlife conservation* (1995), although it is weighted towards a pro-zoo stance.

Although sociologists have had little to say about zoos, a sociological perspective that is useful in debating the zoo controversy is the Sykes & Matza (1957) concept of 'techniques of neutralization'. These authors identified a number of rationalisations which juvenile delinquents used to neutralise social norms. In the case of zoos, the following neutralisations have been made by zoo defenders:

- Denial of responsibility: 'Without zoos, you might as well tell these animals to get stuffed!' (This is a popular slogan – accompanied by a selection of attractive animal portraits – used by zoo defenders to emphasise their responsibility to endangered species; in this case, responsibility is affirmed rather than denied).

- Denial of injury: 'Zoos exist because it's a jungle out there' (This slogan appeared in an advertisement in Geo Australasia 16(1) 1994 and again asserts the role of zoos in preventing rather than causing harm to captive animals).
- Denial of a victim: 'The captive animal has by no means a bad bargain' (Bostock 1993).
- Condemnation of the condemners: This tactic is indicated in a question put to a zoo administrator who implies in her answer that those who condemn zoos are irresponsible.

  Q. *Can zoos survive in times of political correctness and animal rights loonies?*
  A. *Absolutely . . . Remember, there has never been an animal that asked to be put into captivity. Therefore, they are all our responsibility* ('Creature Care', *Herald Sun Weekend*, 28 June 2003, p. 2).

- Appeal to higher loyalties: Rothfels (2002: 175) suggests that the ark metaphor is 'a profoundly resonant justification for [the zoos'] continued existence in the face of their critics'.

This last point is borne out by the sociologist Franklin (1999), who argues that people today in the West want to spend more time with animals, not less; many will have their companion animals and many others will want to see animals in zoos. Franklin, like Wolfe, is highly critical of the animal rights movement for its belief that the appropriate way for humans to treat animals is to leave them alone.

FUTURE DIRECTIONS

Anyone who has read *New Scientist* during the last decade or so could be forgiven for believing the journal's theme was science fiction. The magazine has featured reports of animal patents ranging from cows to fish as well as the onco-mouse which carries a cancer-causing gene that literally makes the animal 'hot property' to medical researchers. Every month, *New Scientist* and other respected technical magazines describe the latest developments in organ transplants, xenotransplantation, transgenic animals, and the genetic engineering of microorganisms, plants and even the prospect of human clones, the subject of a *Time* (Australia) magazine cover story on 8 November 1993. Many people no doubt believe the issue of genetic engineering should be left to scientists as the topic appears to be too technical for non-scientists. The sociologist Barbara Katz Rothman (1995), however, makes a good case for seeing the issue as a social problem rather than a technical or biological one. Looked at in this way, the genetic engineering of animals – and the ethical and social issues it raises – is well within the domain of sociology and, as such, provides an important area of research for postgraduate students in the discipline.

The issue of genetically engineering 'future' animals is conceived by the animal movement – along with vivisection, factory farming, bloodsports and to a lesser extent zoos – as a moral and social problem involving the abuse of animals. Animal liberationists worry that future animals may well be genetically engineered specifically for research, food, hunting and angling, and other purposes such as for entertainment or as companion animals.

Improving Nature? (1996) is the title of a comprehensive account of the science and ethics of genetic engineering by Michael Reiss and Roger Straughan, who argue that each case of genetic engineering has to be decided on its merits. According to them, the genetic engineering of animals to produce life-saving pharmaceuticals – as in the case of Tracey the protein-producing sheep – is a moral necessity; but when animals suffer to satisfy trivial human wants or needs, or when the benefits are uncertain, they oppose it as morally dubious.

Another book to examine the ethical issues associated with the genetic engineering of animals is Bernard Rollin's (1995) evocatively titled *The Frankenstein Syndrome*. Dr Frankenstein's hubris in seeking to create life produces in fictional form a rampaging monster that for many people is emblematic of the folly of modern-day genetic engineering. Rollin argues that the genetic engineering of animals is not intrinsically wrong as many people claim and therefore should not be banned as some environmental philosophers have demanded; he is scathing of the environmentalists' 'new ethic' which fosters the erroneous idea that nature is perfect as it is.

The bottom line for Rollin in genetic engineering is 'the principle of the conservation of welfare', which means that any genetically engineered animals should be no worse off and preferably better off than the parent stock (1995: 179). He concludes that while genetic engineering is not intrinsically wrong, 'it is likely to lead to [greater suffering] in the current exploitive business context' (1995: 181–2). In this sense, he believes the conservation of welfare principle is more relevant to commercial agriculture than to research, though both are closely enmeshed. One could speculate that animals might in the future be genetically engineered to satisfy the needs of scientists, farmers, hunters and anglers as well as consumers. For example, McCarthy & Ellis (1994) have raised the prospect of a transgenic fish which would be better adapted to polluted waters and have qualities such as fighting spirit and size that would be attractive to anglers.

Reiss and Straughan's *Improving Nature?* is much more sceptical about genetic engineering than Rollin's *Frankenstein Syndrome*. Rollin does not believe that an environmental ethic is needed to guide us in addressing the ethical and social issues raised by genetic engineering. Yet his conservation of welfare principle is likely to be ignored when the potential profits from genetically engineered animals are so massive for the biocapitalist. As is currently the case, human greed will conveniently be framed as 'human welfare' when animal protectionists advocate the interests of animals in any genetic revolution in the future. And in dismissing the environmentalists' creed that 'nature cannot be improved', Rollin seems to deny the undeniable, namely the ecological law that governs our relations with nature, which Martell reminds us is a reciprocal relationship involving 'society's effects on nature and the effects of society's impact on nature as they rebound on society' (1994: 24).

A recent illustration of the society–nature relationship can be seen in the circumstances surrounding the outbreak of mad cow disease in the United Kingdom. The Bovine Spongiform Encephalopaphy (BSE) controversy in the UK alerted the world to the farming practice of feeding sheep's brains to cattle, thereby turning a herbivorous animal into a carnivore with fatal consequences for some consumers of beef and the animals themselves. Much more worrying for the consumer of animal flesh, say from the descendants of Tracey the sheep, is the prospect that they could be eating a product containing copies of human genes. Reiss & Straughan (1996) unconvincingly suggest that labelling the product as containing human genes might be the solution but acknowledge that sadly, none of us can be certain of what we are eating given the way the modern food industry operates today. Our only hope is that R.G. Frey's (1983) 'concerned individual', assisted by investigative journalism and the work of activists in social movements as well as critical academic research, will demand an open debate on these life-and-death matters for human and non-human animals alike.

Discussion Questions

1. Are our relations with non-human animals a relevant topic for sociology?
2. Is the term 'animal rights' an oxymoron or an idea whose time has come?
3. Are environmentalists and animal rights supporters kindred spirits or unreliable friends?
4. Which of the following animal rights campaigns would you be most likely to support or challenge (and give your reasons): the campaign against animal experimentation; the campaign against intensive farming; the campaign against recreational hunting?

5. What do you think would be the class/gender composition of the people involved on both sides of the above campaigns?

6. Is it plausible to describe animal rights and vegetarianism as social movements?

7. Discuss the idea that the functions of zoos as suggested by zoo defenders are rationalisations, not rational arguments.

8. Can zoos survive in times of political correctness and animal rights loonies? (This question was posed by journalist Bob Hart to zoo manager Bev Drake in the *Herald Sun Weekend* on 28 June 2003, p. 2.) How would you respond?

9. How can sociologists contribute to an understanding of the issues associated with the genetic engineering of animals?

10. Visit the *Gene Watch* website in the United Kingdom and read their short piece on 'GM animals: Do the Ends Justify the Means?' Have this debate in your class.

Glossary of terms

Anthropocentrism is essentially a human-centred view of the world which privileges humans over other life forms. It is usually contrasted with ecocentrism, which is based on the idea that all life forms are equally important to the ecosystem.

Counter-movements arise in response to the activities of a successful movement which opponents seek to challenge by organising a counter-movement. Thus the animal rights movement's campaigns are countered by groups such as the incurably ill For Animal Research (iiFAR) or the Cattlemen's Association.

Deep/shallow ecology is a dichotomy used to describe two extremes of environmental thinking. Shallow ecologists believe that nature matters, but not as much as humans, while the reverse is true for deep ecologists. This dichotomy corresponds to the rights/welfare division in the animal liberation movement.

Gaia is the name the Greeks gave to their earth goddess. In the present age, radical environmentalists refer to the Gaia hypothesis, which suggests that humans are but one species on the planet with no more of a right to existence than other life forms, which are often more important to the health of the ecosystem.

Nature/'nature' is a term that has been described as among the most difficult in the English language to define. In the present chapter, nature without the quotes is meant to include the natural world on land or sea and in the atmosphere, whereas 'nature' refers to constructed or mediated forms of nature such as nature parks, reserves, zoos, aquariums and the like.

Rights, or more accurately, a rights discourse, is now used by many groups to press their claims. Early social movements such as the Civil Rights movement infused the term with moral and political overtones to make claims on behalf of oppressed humans. We now have a multitude of groups making similar claims on behalf of children, prisoners and even animals. Some people maintain that the concept of animal rights is incoherent since animals, unlike humans, are unable to claim rights.

Social movements: one of the earliest definitions is to see a social movement as a conscious, collective, organised attempt to bring about or resist large-scale societal change by non-institutionalised means. Recent scholars of social movements refer to the importance of a movement's identity and its strategies in framing issues which will appeal to a wide audience.

Social movement organisations (SMOs): these are the organised wing of the broader social movement, for example People for the Ethical Treatment of Animals, or PETA, as a prominent SMO in the broader animal rights movement.

Social problems usually refer to things human societies would be better off without, such as crime, pestilence, war, unemployment, poverty and so on. More recently, racism, sexism, anthropocentrism and speciesism have been added to the list. These social problems have been the focus of social movements which campaign against the relevant 'ism'. In the case of animal liberation, the 'isms' are seen as interconnected systems of oppression.

Speciesism refers to the practice of discriminating against non-human animals because they are perceived as inferior to the human species, in much the same way that sexism and racism involve prejudice and discrimination against women and people of different colour.

References

Adams, C. 1990, *The Sexual Politics of Meat: A feminist-vegetarian critical theory*, Cambridge: Polity Press.

Arluke, A., and J. Groves 1998, 'Pushing the boundaries: scientists in the public arena'. In L. Hart (ed.) *Responsible Conduct in Research*, Oxford/ New York: Oxford University Press.

Bostock, S. 1993, *Zoos and Animal Rights: The ethics of keeping animals*, London: Routledge.

Cartmill, M. 1993, *A View to a Death in the Morning: Hunting and nature through history*, Cambridge MA: Harvard University Press.

De Courcy, C. 1995, *The Zoo Story: The animals, the history, the people*, Melbourne: Penguin Books.

Dizard, J.E. 1994, *Going Wild: Hunting, animal rights, and the contested meaning of nature*, Amherst MA: University of Massachusetts Press.

Elias, N. 1978, *The Civilising Process*, vol. 1, *The History of Manners*, Oxford: Basil Blackwell.

Fiddes, M. 1991, *Meat: A natural symbol*, London: Routledge.

Franklin, A. 1999, *Animals and Modern Cultures: A sociology of human–animal relations in modernity*, London: Sage.

Frey, R.G. 1983, *Rights, Killing and Suffering: Moral vegetarianism and applied ethics*, Oxford: Blackwell.

Groves, J. 1997, *Hearts and Minds: The controversy over laboratory animals*, Philadelphia PA: Temple University Press.

Hargrove, E. 1992, *The Animal Rights/Environmental Debate: The environmental perspective*, State University of New York Press.

Jasper, J., and D. Nelkin 1992, *The Animal Rights Crusade: The growth of a moral protest*, New York: Free Press.

LaFollette, H., and N. Shanks 1996, *Brute Science: Dilemmas of animal experimentation*, London: Routledge.

MacCulloch, J. 1993, Creatures of Culture: The Animal Protection and Preservation Movements in Sydney, 1880–1930, unpublished PhD thesis, University of Sydney.

Malamud, R. 1998, *Reading Zoos: Representations of animals and captivity*, New York University Press.

Martell, L. 1994, *Ecology and Society: An introduction*, Cambridge: Polity Press.

McCarthy, C., and G. Ellis 1994, 'Philosophic and ethical challenges of animal biotechnology', *Hastings Center Report* 24: 14–30.

Mukerjee, M. 1997, 'Trends in Animal Research in a Forum on the Benefits and Ethics of Animal Research', *Scientific American* February: 79–93.

Mullin, B., and G. Marvin 1987, *Zoo Culture*, London: Weidenfeld & Nicolson.

Munro, L. 2001, *Compassionate Beasts: The quest for animal rights*, Westport CT: Praeger.

Norton, B., M. Hutchins, E. Stevens and T. Maple 1995, *Ethics of the Ark: Zoos, animal welfare and wildlife conservation*, Washington/London: Smithsonian Institution Press.

Penman, D. 1996, *The Price of Meat*, London: Gollancz.

Plous, S. 1993, 'The role of animals in human society', *Journal of Social Issues* 49(1).

Reiss, M.J., and R. Straughan 1996, *Improving Nature? The science and ethics of genetic engineering*, Cambridge University Press.

Rifkin, J. 1992, *Beyond Belief: The rise and fall of the cattle culture*, New York: A Dutton Book.

Robinson, M. 1996, Foreword to R.F. Hoage and W.A. Deiss (eds) *New Worlds, New Animals: From menagerie to zoological park in the nineteenth century*, Baltimore MD/London: Johns Hopkins University Press.

Rollin B. 1995, *The Frankenstein Syndrome: Ethical and social issues in the genetic engineering of animals*, Cambridge University Press.

Rothfels, N. 2002, *Savages and Beasts: The birth of the modern zoo*, Baltimore MD/London: Johns Hopkins University Press.

Rothman, B.K. 1995, 'Of maps and imagination: sociology confronts the genome', *Social Problems* 42: 1–10.

Sykes, G., and D. Matza 1957, 'Techniques of neutralization: a theory of delinquency', *American Sociological Review* 22: 664–70.

Turner, E.S. 1992, *All Heaven in a Rage*, Fontwell, UK: Centaur Press.

Vidal, J. 1997, *McLibel: Burger culture on trial*, London: Macmillan.

Wolfe, A. 1993, *The Human Difference: Animals, computers and the necessity of social science*, Berkeley CA: University of California Press.

CHAPTER FIVE

GOVERNING ENVIRONMENTAL HARMS IN A RISK SOCIETY

Vaughan Higgins and Kristin Natalier

This chapter critically examines the 'risk society' thesis and the questions it raises concerning the definition and governing of environmental harms. We first outline the main features of Beck's approach to risk. The limitations of this work are then discussed drawing on social constructionist concepts to highlight the main areas of debate. Finally, we provide a critique of both realist and social constructionist perspectives on risk using the post-structuralist analytical frameworks of governmentality and actor-network theory. We argue that these two latter frameworks represent a coherent way of 1) moving beyond the dualistic objectivist/subjectivist thinking that characterises both realism and social constructionism, and 2) demonstrating the complex ways in which environmental harms are co-constructed as risks.

BACKGROUND TO THE ISSUES

To understand how environmental harms assume prominence as risks it is useful to consider first Ulrich Beck's seminal work, *Risk Society: Towards a new modernity* (1992). While Beck was by no means the first sociologist to write about the nature and management of risk, his work has had a major influence in environmental sociology and therefore provides a starting point in exploring the sociological literature on the production, definition and responses to ecological harms.

Emergence of the risk society

Beck's *Risk Society* is a theory of modernisation. It describes a shift from classical to reflexive modernisation. For Beck, classical modernisation is characterised by a politics centred on material progress and the distribution of wealth and prosperity ('goods'). But the taken-for-granted assumptions of progress have recently been displaced by a concern with the negative environmental consequences of development ('bads'). Risk has become the organising principle of late modernity: individuals and institutions focus on the management, allotment and avoidance of potential danger. These processes are part of other substantial historical shifts: individualisation; the declining significance of class; the rise of new social movements and the critique of expert knowledge.

By Beck's (1992: 21) definition, risk is 'a systematic way of dealing with hazards and insecurities induced and introduced by modernization itself'. In other words, risk is a consequence of societal attempts to intervene in and control environmental hazards. It demands a constant engagement with the future and in so doing, it shapes the present: it is 'something which has not happened yet, which frightens people in the present and therefore they might take action against it. Risk is not catastrophe; if catastrophe happens it is a fact, an event' (Boyne 2001: 57). As such, the risk society is not defined by oil spills or nuclear meltdowns; it is defined by their possibility, and how we might, as a society and as individuals, respond to this potential.

Beck draws a sharp distinction between incalculable hazards and calculable risks. People have always faced hazards such as illness, death and social upheaval. But the risks of late modernity – Beck uses global warming, the greenhouse effect and the thinning of the ozone layer as exemplars – are unique. Contemporary risks are set apart through:

- *Origins*: Danger and disaster once struck – and were avoided – through fate, or the will of 'gods, demons or Nature' (Beck 1992: 98). In late modernity, risks are generated by society. They arise through the failure of social institutions to control the risks that are inherent to them. Chemical spills or radiation poisoning are more than a by-product of industrialism and capitalism; they are consistent with the logic of industrial capitalism and modernisation.

- *Scope and effect*: The world today faces the possibility of apocalypse. Even those risks that fall short of the complete annihilation of humankind, or the environment, display an unprecedented reach. They transcend time: their effects are not limited to present generations, and indeed may only be fully experienced by people who have not yet been born. These risks are global, extending far beyond the surrounds of any one factory that might produce them; they cannot be limited to the territory of a particular state. Due to the 'boomerang effect', those who generate risks cannot export them elsewhere and escape them – the threat will rebound. Rich people are no more able to protect themselves from some of the key dangers that define the modern world than are the poor: 'Nuclear contamination is egalitarian and in a sense "democratic". Nitrates in the groundwater do not stop at the director general's tap' (Beck 1992: 109). Further, the risks have the potential to 'induce systematic and irreversible harm' (Beck 1992: 22). This harm is often incalculable and can be neither insured against nor compensated for: there is no possibility of returning a person, culture or place to the state it was in before the event occurred. In short, instrumentality and rational control, fundamental organising principles of modernity, have been undermined in the emerging society.

- *The difficulties of identification:* Smoke billowing from factory stacks was once a visible indication of the destructive environmental effects of development. In effect, no specialist knowledge was needed to identify them. Today, people cannot rely on their senses to foresee or avoid danger. A 'cloud of radiation' does not look like a cloud, nor can we look up into the sky, see the 'ozone hole', and walk around it. Risks are invisible, they cannot be smelt, heard, touched or tasted; for Beck, they are 'unknowable' to lay people.

The role of expertise

Given that risks are undetectable, people are reliant on experts. This is an ambiguous and ambivalent relationship. On the one hand, risks are not detectable by human senses and so science is necessary in order to identify the existence of risk. On the other hand, the legitimacy of science is increasingly challenged. The rules and proofs of science are at odds with the incalculable nature of risk, its systemic sources and its global and temporal reach. It can be difficult to establish a relationship between an adverse health or environmental outcome and the practices of any one individual, factory or corporation, at least to a level that meets recognised scientific, legal or statistical standards of proof. This has the effect of rendering monitoring and prosecution difficult. However, the demands of the institutions developed in simple modernity do not fit with people's lived experiences as 'people themselves become small, private alternative experts in risks of modernization. For them, risks are not risks, but pitifully suffering, screaming children turning blue . . . The "blank spots" or modernization risks, which remain "unseen" and "unproven" for the experts, very quickly take form under their cognitive approach' (Beck 1992: 61).

In Beck's risk society, expert knowledge is relied on, critiqued and appropriated in 'a dialectic of expertise and counter-expertise' (Beck 1992: 30).

Redefinition of social progress

According to Beck, new risks are the logical endpoint of social, capitalist and industrial processes. In simple modernity, people believed
in progress and saw its achievement through the development of more
efficient markets, advanced science, and appropriate technology. Any
problems in the environment would be addressed through the very science and technology that contributed to them in the first place. Beck
terms this 'organised irresponsibility' whereby society endangers itself
by failing to acknowledge and address the source of risk in its own social
and institutional systems. The few who sounded warnings were labelled
retrograde nay-sayers or fearmongers.

However, the consequences of a society built on science and industry are no longer passing unnoticed: a reflexive risk society – more
commonly referred to as 'reflexive modernization' in Beck's text – has
emerged. People and institutions are suspicious of industries and earlier modes of production and management (Boyne 2001: 58). This creates new political alignments. Class has lost much of its relevance in
a post-scarcity society, whereby most of our material needs are now
met. Any lingering inequalities are reframed with reference to the
strategies, successes and failures of individuals – they are no longer
ascribed on a group level. A consequence is that an environmental 'sub-
politics' emerges where people organise collectively around new risks
rather than previous socio-economic allegiances. These concerns build
coalitions between individuals who may not otherwise have worked
together. Actions are no longer bound to traditional political processes
and people's concerns are no longer centred purely on the ideals of
progress – either techno-scientific or social. In a risk society, unprecedented risks are interlocked with significant social change.

KEY DEBATES

Risk Society has become one of the pre-eminent sociological texts; its
sweeping scope and impassioned critique of environmental crises have
struck a chord with the public and sociologists alike. However, academic engagement with this work has revealed a series of shortcomings.
The following section explores the debates associated with this concept,
and uses the criticisms of Beck's work as a starting point in exploring
some of the broader tensions between the dominant perspectives on
environmental risk: realism and social constructionism.

Realist versus constructed notions of risk

Although his conceptualisation is not always clear, Beck is often
criticised for adopting a realist definition of risk which argues that

environmental threats and their material outcomes exist independently of social perception or cultural interpretation. As Dean (1999: 136) comments, Beck assumes that risk is such a central feature of modern existence because real riskiness has increased and has 'escaped the mechanisms of its calculation and control'. For Beck, these effects are real, existing irrespective of the absence of scientifically mandated relationships or legal definitions of causation. While this demonstrates the serious and concrete nature of the risks faced by modern societies, scholars such as Elliot (2002: 300–1) note that Beck's theory is overly rationalist and neglects the subjective ways in which risks are constructed; in sum, 'it cannot grasp the hermeneutical, aesthetic, psychological and culturally bounded forms of subjectivity and intersubjectivity in and through which risk is constructed and perceived'. These comments are consistent with a social constructionist perspective that focuses on the interpretive frameworks, cultural assumptions and forms of contestation that create or conceal environmental problems and issues.

According to a social constructionist approach, risks are socially defined and created rather than simply objective phenomena waiting to be discovered. Hannigan (1995: 96–100), drawing on the work of Hilgartner, argues that there are three major conceptual elements in the social definition of risk:

- An *object deemed to pose a risk*: To assume significance as a risk, an object needs to be defined as such. Often there may be competing claims over the nature of the object as risky.
- A *putative harm*: This involves debate over the harmful effect of the risk in question. Again, the harm involved can be the subject of contestation and debate. For example, forest fires might be seen as a cause of destruction, but can also be viewed by ecologists as a natural means of renewing woodlands.
- A *linkage alleging some causal relationship between the object and harm*: Scientific and statistical evidence is frequently drawn on to 'prove' a link between the object and harmful (or lack of harmful) effects. However, the layers of proof are multiple and include not only scientific but also legal and moral claims. As a consequence, groups, such as those opposed to genetically modified foods, have been able to use moral arguments to mobilise public opinion and contest what they see as tampering with nature.

Nevertheless, a social definition of risk has been criticised by realists such as Dunlap & Catton (1994), who argue that it fails to say much about the 'reality' of environmental harms. In their view,

environmental sociology has an obligation to address ecological problems in practical rather than deconstructive terms; as an intellectual
approach, social constructionism provides no basis to do so (see also
Benton 1994; Martell 1994). However, Burningham & Cooper (1999)
point out that the debate between realism and constructivism is centred
on a straw person. It is not social constructionism per se that is attacked,
but rather 'strong social constructionism', which does not allow for the
existence of a reality outside discourse or cultural definitions. Much
of the work in environmental sociology could better be termed 'mild
or contextual constructionism'. This approach, advocated by sociologists such as Hannigan (1995) and Capek (1993), focuses on the ways
in which people interpret environmental problems, highlighting the
contested nature of particular claims, but proceeds on the assumption
that material and often serious risks exist independently of any group's
claims. Nor is a realist approach so strict that it shares no ground with
constructionism; the changing and contested nature of scientific knowledge, in particular, is regarded as relatively problematic in sociology
generally. Thus Irwin (2001) suggests that it is more accurate to talk
of 'constructive realists' and 'real constructivists'.

Risk society versus socially constructed risks

The divisions between 'lay' and 'expert' or 'scientific' knowledge has
created a further area of debate. Within Beck's reflexive modernisation
thesis, debates over risk occur between experts and counter-experts;
any non-institutional critique derives from personal experience. Wynne
(1996) argues that this dichotomy demands reconsideration. His
account of the interactions between English sheep farmers and scientists in the wake of the 1986 Chernobyl nuclear accident in the Ukraine
breaks down the expert/lay divide, highlighting the accurate, detailed
and contextual knowledge of the farmers (see also Murdoch & Clark
1994). This study stands as a reminder that an absence of formal cre-
dentials does not necessarily mean that 'local' knowledge is 'epistemi-
cally vacuous' (Wynne 1996: 61). In fact, while the knowledge of the
farmers concerning the impact of radioactive contamination on their
farms was considered by the scientists as lacking precision and calcu-
lability, this knowledge showed a detailed understanding of the local
environment. The failure of the scientists to take account of the farmers' expertise resulted in direct conflict 'over the appropriate design of
scientific experiments' (Wynne 1996: 67).

The relationship between expert and lay knowledge (these terms
will remain for the sake of convenience) generates a further point of

contestation: the role of trust. Beck (1992) presents an ambivalent relationship between dependence and mistrust. The invisible nature of risk means that knowledge must be mediated through science, but reflexive modernisation – and personal experience – has highlighted its systemic failings. Anthony Giddens (1990) adopts a similar theme, but concludes that science, along with other 'expert systems', remains a fundamental source of ontological security in a world that is impenetrable to those without specialised knowledge. For these authors, trust is predicated on the accuracy of the information provided by experts; it is rational and calculative and ultimately dependent on formal science; it breaks down when the experts obviously (sometimes disastrously) get things wrong. In contrast, Wynne (1996) argues that a belief in expert knowledge is predicated on more than simple accuracy. Risk is relational: definitions and calculations of risk are rooted in connections between people and institutions. When people assess risk, they are not only responding to the risks attached to events or technologies and the quantifiable harm they might cause; they also consider the behaviours and responsibilities of the relevant institutions. These judgements cannot be made independently of pre-existing relationships. When communities are dependent on institutions, overt and publicly expressed mistrust problematises these social relationships. For Wynne, risk is as much about social identities and social relationships as it is about the calculable and identifiable potential of a negative outcome.

Sub-politics versus power and domination

A third criticism of Beck's work is that he neglects patterns of power and domination in the structuring of risk (see Hannigan 1995: 103–7; Elliot 2002: 302–6). According to Beck, the risk society is characterised by an 'equalising effect' (Beck 1992: 35) where both rich and poor face the same dangers. The proliferation of global risks means that not even the wealthy can escape their effects. This, in conjunction with the public loss of trust in science to address these problems, results in a reinvention of politics in which there is a shift in focus from institutionalised politics towards sub-political forms of active self-management. Beck argues that a reflexive sub-politics has the greatest capacity to control the negative consequences of risk. However, Beck's work tends to assume uniformity, and that 'reflexive modernization' transforms previous social divisions and relations of power and domination into a single politics of risk. What is needed, according to Elliot (2002: 306), is the development of 'methods of analysis for explicating how patterns of power and domination feed into, and are reconstituted by, the socio-symbolic structuring of risk'.

A social constructionist approach addresses this challenge by examining the role of power relations in framing risk issues. It investigates the competing frames through which debates over risk are played out, and the context-dependent nature of risk understandings. Of particular significance is how risk professionals are able to present their views as rational and objective, thereby representing competing frames as 'irrational', 'emotional' and 'non-scientific' (Hannigan 1995: 104–5). From this perspective, it is necessary to examine claims by scientists over the nature of risk, the counter-claims made by citizen groups, non-government organisations and social movements, and how this process contributes to 'popular concerns and risk frames [to be] subordinated to those which are preferred by the powerful in society' (Hannigan 1995: 106). Of course, this is not to suggest that 'scientific knowledge' always triumphs over popular understandings of risk. As Irwin (2001) argues, 'environmental knowledge is complex, problematic and characterised by uncertainties and ambivalences'. This implies caution is required in attributing unity to the claims of various groups since the knowledge on which these claims is made is always contested and changeable.

Social constructionist approaches provide a valuable counterpoint to the realism that is argued by many scholars to characterise Beck's theory of risk. In particular, social constructionism draws attention to the broader social and cultural context in which claims and counter-claims regarding risk and riskiness are constructed, and the relationship between 'expert' and 'lay' knowledge in this process. However, this 'culturalist' approach has two problems according to Crook (1999: 176–7). First, risk identification, assessment and management are approached in the same way as other types of claims-making. There is no thoroughgoing appreciation of their specificity, of the ways in which they may differ from other issues that are subject to social construction. As a consequence, questions of how risks assume prominence as objects of knowledge, and the specific practices that define, order, and give durability to certain risks over others are obscured as part of general claims-making processes. Second, the approach presumes that despite the diversity of definitions and viewpoints, the cultural and social systems from which they originate are pure or homogeneous. Thus despite the recognition of the plurality of claims and competing viewpoints in environmental controversies, the construction of risk is based on unitary notions of culture and 'the social'. In effect, this neglects the complex ways in which risks are produced, and governed, as 'natural' or 'social' phenomena.

Risk as ordering: a post-structuralist approach

A governmentality approach focuses on how environmental harms are rendered knowable and calculable as objects of knowledge. Thus risk represents a particular *style* of thinking that entails new ways of understanding and acting on harms (Rose 1999: 246). While this approach has had a significant impact on the social sciences in the past fifteen years, it has yet to be applied to the study of environmental risk in a systematic way. It therefore shows much potential for addressing the problems raised by realist and social constructionist perspectives. A governmentality approach argues that:

> There is no such thing as risk in reality. Risk is a way – or rather, a set of different ways – of ordering reality, of rendering it into a calculable form. It is a way of representing events in a certain form so that they might be made governable in particular ways, with particular techniques and for particular goals. It is a component of diverse forms of calculative rationality for governing the conduct of individuals, collectivities and populations. (Dean 1999: 131)

In other words, far from existing prior to attempts to manage it, risk is an effect of specific strategies of governing. Social constructionists agree that the management of risk needs to be examined in its social context. However, they treat risk as an unproblematic object of different groups' beliefs, values and interests. In effect, even though there may be a variety of claims and counter-claims made over the causes and consequences of harm, there is usually mutual agreement over the risk object (see Hannigan 1995: 98). A governmentality approach problematises this by arguing that seemingly inconsequential attempts to name and manage environmental harms actually define and redefine the discursive boundaries of so-called risky events and behaviour. This is not to suggest risk is simply a discursive construction, as is often claimed by critics of 'postmodern' theory (see Barry 1999), but that the 'reality' of risk is defined and shaped through attempts to render it knowable. As Dean (1999: 131) continues, 'what is important about risk is not risk itself, but the forms of knowledge that make it thinkable . . . the techniques that discover it . . . the social technologies that seek to govern it . . . and the political rationalities and programmes that deploy it'. Risk then is an outcome of attempts to 'know' environmental harm, to describe its features, to calculate its potential effects and costs, to write about, and graphically represent it, to *order* these harms.

Crucial to this process are the techniques of expertise that provide 'a kind of *intellectual machinery* for government', depicting harm 'in a way which both grasps its truth and re-presents it in a form in which it can enter conscious political calculation' (Rose & Miller 1992: 182). In other words, before an environmental harm such as, for example, acid rain, can be addressed as a risk, it needs to be represented and rendered 'knowable' as an object of knowledge that has certain characteristics, regularities and causes. This is made possible through what Miller & Rose (1990), drawing on the work of Latour (1987), call inscription devices (e.g. tables, graphs, academic articles) that accumulate and order harms in a technical form and thereby enable authorities to act on distant events, places and people. Thus these devices are more than neutral means of recording already-existing risks, but material techniques that transform environmental harms into definable, diagnosable and governable issues.

This means that the definition and governing of an environmental harm as a risk is a contingent effect of how experts attempt to negotiate and align other actors with expert-defined representations of risk. To understand how these alliances are forged, we now turn to the analytical insights of actor-network theory, and specifically the early work of French sociologist Bruno Latour.

For Latour, and other adherents of the approach termed loosely actor-network theory, risks need to be viewed as contingent outcomes of attempts at network-building, whereby scientists build representations of harms and attempt to enrol other groups to actively support these representations. This process – termed *translation* – focuses on the extension of scientific networks, the work involved in aligning the goals of others with that of the scientists, and the inscriptions deployed (such as charts and graphs) to stabilise these associations (see Callon 1986; Latour 1986, 1987).

> If successful, translation enables an actor, or group of actors (such as scientists), to speak on behalf of others. As a consequence, an environmental harm previously the subject of dispute comes to be stabilised as a universal 'fact' – a risk whose characteristics and consequences are no longer subject to dispute. However, it is important to note that translation is not a linear process where an instrumental scientific rationality progressively dominates, or where 'the dominant rationality which comes from the risk establishment is superimposed over the popular frame due to a power differential'. (Hannigan 1995: 104)

It is, in fact, a precarious process that relies on the alignment of often disparate actors (Star 1991; Singleton & Michael 1993; Clark & Murdoch 1997). The work of translation is characterised by ambivalence, complexity, multivocality and competing network-building activities that shape, and may undermine, attempts by scientific 'experts' to define and manage risk.

A significant aspect of a translation approach is that it questions the common tendency to see risks as either objectively real or socially constructed. Rather than resorting to 'nature' or 'society' as the basis for explaining risk, Latour (1993) believes it is better to dispense with such dualisms and to explore how these two domains are produced as a consequence of attempts at network-building. Thus this perspective focuses on the *co-construction* of the social and natural (Irwin 2001: ch. 7). For Latour, modernity is based on the production of what he calls 'hybrids', or mixtures of nature and culture. At the same time, however, this hybridity is concealed by practices of 'purification' that 'create two entirely distinct ontological zones: that of human beings on the one hand; that of nonhumans on the other' (Latour 1993: 11). While hybrid networks of nature–culture are constantly being created, purification constructs these domains as two very different categories: 'nature' is the legitimate domain of natural scientists while 'culture' is what social scientists are seen to study. According to Latour, this dualistic thinking limits a more complex understanding of the network-building activities (translations) through which harms are constituted, stabilised and 'purified' as 'human' or 'non-human' risks. Practices of purification *and* translation therefore need to be studied together in order to understand how (hybrid) risks are produced and the effects in terms of how these risks are categorised as 'natural', 'social', 'moral' or 'political' problems. We should not assume an *a priori* distinction between these categories since they are outcomes of environmental controversies. In other words it is 'only when networks have been established, and roles and identities distributed within them, that a clear-cut difference emerges between "things out there" and "humans in here"' (Murdoch 1997: 744).

This has clear relevance to controversies over environmental harm. For example, the spilling of toxic waste might be framed simultaneously as an issue concerning nature (it has a negative effect on ecosystems), a social issue (waste as a harmful consequence of capitalist production), a moral issue (big business not taking due care with the management of waste) or a political issue (issues over the regulation of industry). The key insight of a translation approach is that rather than resorting to one or other of these explanations, it is of more use to trace the translations (and purification) that enable such categories to be hardened into indisputable facts, and the configuration(s) of risk to which this gives rise.

FUTURE DIRECTIONS

This chapter argues that the notion of the 'risk society' has been influential in shaping sociological debate on the definition and management of risk. However, its realist assumptions have made it the subject of continued debate. From a social constructionist perspective, Beck's arguments fail to account for the multiple, and often competing, ways in which environmental harms are constituted as risks; the arenas of risk construction in which environmental controversies are played out; and the forms of power and domination that frame the definition and governing of harms. The criticisms of Beck's work draw attention to broader tensions between 'realist' and 'social constructionist' risk perspectives.

Social constructionism has proved useful in showing the social and cultural context in which risks are defined, managed and contested. Nevertheless, we have argued that it has little to say concerning the actual practices of risk identification, and how these are deployed in co-constructing actor-networks that give durability to particular representations of risk. According to Irwin (2001: 179), this co-constructionist approach represents the most useful way forward in raising 'fresh possibilities for sociological analysis'; in particular, it 'forces us to re-evaluate not only the usage of natural arguments in environmental debates but also the shifting definition of the social' (Irwin 2001: 173). While co-construction has clear merits in examining how risks are constituted, it is also important to acknowledge its limitations. For many critics, co-constructionism is too relativist, blurring the distinction between 'human' and 'non-human'. As a result, some claim that this neglects human responsibility for environmental harm. For instance, Murdoch (2000: 129) warns that 'the price for adopting this approach may be a diminished ability to understand "human exemptionalism" so that it becomes difficult to adjudicate on a specifically human propensity to both cause and remedy environmental destruction'. The ontological status of 'human' and 'non-human' is likely to make the concept of risk an ongoing future site of debate among environmental sociologists.

Discussion Questions

1. How does Beck's *Risk Society* thesis conceptualise the nature and governing of environmental risk?
2. What are the main merits and limitations of the *Risk Society* thesis? Why?
3. Are the risks we face today fundamentally different from those of past epochs?
4. Is there such a thing as an incalculable risk?

5. In what ways is risk socially constructed?

6. What is the relationship between expert and lay knowledge in the defini-tion and management of risk?

7. What are the key differences between realist, social constructionist and post-structuralist approaches to the definition and governance of risk?

8. Do you agree with Dean (1999: 131) that risk is a means of ordering reality? Why or why not?

9. Think of an environmental controversy with which you are familiar. In what ways are different rationalities of risk, and risk management, evident?

Glossary of Terms

Co-Constructionism: a sociological perspective that conceptualises 'natural' and 'social' phenomena, or 'real' and 'socially constructed' risks, as outcomes or effects of environmental controversies. Rather than treating nature and culture as two separate domains, this approach explores the work of *translation* (see below) involved in producing such distinctions.

Governmentality: a sociological perspective that views risk as configurations of rationalities and techniques for ordering harms. The 'reality' of risk is defined and governed by attempts to write about, represent and calculate various harms.

Realism: a powerful approach in environmental sociology which argues that the materiality and effects of risk exist independently of, and externally to, people's attempts to understand and manage it. That is to say, risk pre-exists human attempts to define and manage environmental harms.

Reflexive modernisation: a historical shift whereby the taken-for-granted prin-ciples and practices of industrial society are confronted in late modernity. Individuals and society as a whole apply an increasingly sceptical scrutiny to modernist projects and outcomes, which are reinterpreted as problems.

Risk refers to the possibility of particular outcomes. In simple modernity, risks were believed to be calculable, and thus manageable. Risks are not hazards, which can be neither calculated nor controlled.

Risk Society: a negative consequence of modernisation in which risk assumes a central place in the ordering of society. The risks faced in this society are qualitatively different from those of the past, and their difference forms the basis for reconsidering relationships, institutions and practices.

Social constructionism: in the context of environmental issues, social con-structionism is an umbrella term that emphasises the cultural definitions and interpretive frameworks that give shape to particular environmental issues as social problems.

Sub-politics: a new form of politics, developing in response to the risks now faced in late modernity. This is not based on traditional political systems, party loyalties or class divides but rather seeks to change things via alternative forms of action.

Translation: a process whereby an actor, or group of actors, attempts to enrol and align others into its representation of risk and thereby be conferred with the authority to speak on behalf of these other actors. From this perspective, society or nature should never be seen as a way of explaining the nature of risk; rather, these domains are a consequence of the alignment and stabilisation of heterogeneous associations of actors and resources.

References

Adam, B., and J. Van Loon 2000, 'Introduction: repositioning risk: the challenge for social theory'. In B. Adam, U. Beck and J. Van Loon (eds) *The Risk Society and Beyond: Critical issues for social theory*, London: Sage, pp. 1–31.

Barry, J. 1999, *Environment and Social Theory*, London: Routledge.

Beck, U. 1992, *Risk Society: Towards a new modernity*, transl. M. Ritter, London: Sage.

Benton, T. 1994, 'Biology and social theory in the environmental debate'. In M. Redclift and T. Benton (eds) *Social Theory and the Global Environment*, London: Routledge, pp. 28–50.

Boyne, R. 2001, 'Cosmopolis and risk. A conversation with Ulrich Beck', *Theory, Culture and Society* 18: 47–63.

Burningham, K., and G. Cooper 1999, 'Being constructive: social constructionism and the environment', *Sociology* 33(2): 297–316.

Callon, M. 1986, 'Some elements of a sociology of translation: domestication of the scallops and the fishermen of St Brieuc Bay'. In Law, *Power, Action and Belief*, pp. 196–233.

Capek, S. 1993, 'The "environmental justice" frame: a conceptual discussion and an application', *Social Problems* 40: 5–24.

Clark, J., and J. Murdoch 1997, 'Local knowledge and the precarious extension of scientific networks: a reflection on three case studies', *Sociologia Ruralis* 37(1): 38–60.

Crook, S. 1999, 'Ordering risks'. In Lupton, *Risk and Sociocultural Theory*, pp. 160–85.

Dean, M. 1999, 'Risk, calculable and incalculable'. In Lupton, *Risk and Sociocultural Theory*, pp. 131–59.

Dunlap, R., and W.R. Catton 1994, 'Struggling with human exceptionalism: the rise, decline and revitalisation of environmental sociology', *American Sociologist* Spring: 5–30.

Elliot, A. 2002, 'Beck's sociology of risk: a critical assessment', *Sociology* 36(2): 293–315.

Giddens, A. 1990, *The Consequences of Modernity*, Stanford University Press.

Hannigan, J.A. 1995, *Environmental Sociology: A social constructionist perspective*, London: Routledge.

Irwin, A. 2001, *Sociology and the Environment*, Cambridge: Polity Press.

Latour, B. 1986, 'The powers of association'. In Law, *Power, Action and Belief*, pp. 264–80.

Latour, B. 1987, *Science in Action: How to follow scientists and engineers through society*, Cambridge MA: Harvard University Press.

Latour, B. 1993, *We Have Never Been Modern*, transl. Catherine Porter, London: Harvester Wheatsheaf.

Law, J. (ed.) 1986, *Power, Action and Belief: A new sociology of knowledge?* London: Routledge & Kegan Paul.

Lupton, D. (ed.) 1999, *Risk and Sociocultural Theory: New directions and perspectives*, Cambridge University Press.

Martell, L. 1994, *Ecology and Society: An introduction*, Cambridge: Polity Press.

Miller, P., and N. Rose 1990, 'Governing economic life', *Economy and Society* 19(1): 1–31.

Murdoch, J. 1997, 'Inhuman/nonhuman/human: actor-network theory and the prospects for a nondualistic and symmetrical perspective on nature and society', *Environment and Planning D: Society and Space* 15: 731–56.

Murdoch, J. 2000, 'Ecologising sociology: actor-network theory, co-constructionism and the problem of human exemptionalism', *Sociology* 35(1): 111–33.

Murdoch, J., and J. Clark 1994, 'Sustainable knowledge', *Geoforum* 25(2): 115–32.

Rose, N. 1999, *Powers of Freedom: Reframing political thought*, Cambridge University Press.

Rose, N., and P. Miller 1992, 'Political power beyond the state: problematics of government', *British Journal of Sociology* 43(2): 173–205.

Singleton, V., and M. Michael 1993, 'Actor-networks and ambivalence: general practitioners in the UK Cervical Screening Programme', *Social Studies of Science* 23: 227–64.

Star, S.L. 1991, 'Power, technology and the phenomenology of conventions: on being allergic to onions'. In J. Law (ed.) *A Sociology of Monsters: Essays on power, technology and domination*, London: Routledge, pp. 26–56.

Wynne, B. 1996, 'May the sheep safely graze? A reflexive view of the expert/lay knowledge divide'. In S. Lash, B. Szerszynski and B. Wynne (eds), *Risk, Environment and Modernity: Towards a new ecology*, London: Sage, pp. 44–88.

CHAPTER SIX

WHEN THE POPULATION CLOCK STOPS TICKING

Natalie Jackson

Three main themes have long characterised the debate on population and the environment. These are: the neo-Malthusian argument that population growth is 'bad' for the environment; the Cornucopian argument (sometimes called the Nationalist, Boomster or Tech-fix argument) that population growth is 'good' for the environment; and the Marxian/Socialist argument that population growth is not directly related to environmental impact but that many economic and political factors intervene between the two. Common to each is the idea that population growth and its continuance are a given, and closely related are two burning questions: whether there should be less, more, or stable numbers of people, both globally and in individual countries, and whether governments should employ population and other policy interventions to achieve these objectives.

Answers to these questions are readily forthcoming – and almost always reflect the theoretical or ideological perspective(s) of their proponents. 'Deep Green' environmentalists, for example, tend to argue from the neo-Malthusian position, that there are too many people, and that fewer are needed if major environmental and social catastrophes are to be avoided. In contrast, business councils and related interest groups tend to argue from the Cornucopian perspective, that there are too few people, and more are needed to facilitate wealth creation and the economies of scale that would deliver less environmentally damaging practices and technologies; while those concerned with social justice tend to argue from the Marxian/Socialist, 'Red Green' point of view that population size is a moral and distributional issue, less to do with how many people a country should have than with the extent to which those people – and their governments – could or should alter their values and consumption habits in order to achieve a more globally equitable and sustainable future.

This chapter elaborates these arguments. But it also weaves in a new perspective, positing it as a 'fourth pillar' for the population–environment debate. This is the proposition that, over the long term, population dynamics (birth and death rates) contain within them an internal logic of growth and decline, and that the environmental impact of population could follow a similar trajectory *if* these dynamics were more broadly understood and appropriately acted on.

Underlying the proposition is the argument that, in the population–environment literature, the 'population variable' has typically been taken to be *growth in the size* of global or national populations. There has been minimal engagement with either the underlying cause of that growth – the phenomenon known as the *demographic transition* (the global decline from high to low birth and death rates) – or the related changes in *age composition* (the proportions at each age) that will eventually – within a century – almost certainly bring about zero growth and/or global population decline (e.g. Lutz et al. 2001).

The deficit means that the population–environment debate is missing a vital piece of evidence on which resolutions to 'the problem' might be more adequately developed. If, for example, it was more widely understood that the world's population is going to increase by some 3 billion over the next half-century, but that numbers will then level off and begin to decline, then those in the developed countries might feel more inclined to invest more resources in this arguably short-term challenge. Equally, if it was more widely understood that the cause of the yet-to-be-realised growth (all of which will occur in the developing countries) is not high birth rates at all – which are everywhere falling – but rather that the infant mortality rates of the developing countries are simply declining ahead of their birth rates, those resources might be more appropriately invested.

Projected time horizons to the onset of peak population numbers and global decline do in fact vary widely (Lutz 1996; Bongaarts & Bulatao 2000; Lutz et al. 2001; Wilson 2001a), but few demographers question the overall trend. Most developed countries are expected to enter natural decline during the next few decades (United Nations 2000), along with China and many more recently developed and still developing Asian countries (Lutz et al. 2003). Most remaining countries are expected to follow suit during the second half of the century, with the global population reaching zero growth and/or entering natural decline around 2100.

Despite the somewhat provisional nature of these trends (Wilson 2001b), the importance of incorporating them into the population–environment debate cannot be overstated. As sociologists Thomas & Thomas (1928: 572) cautioned: if men [sic] define situations as real, they will be real in their consequences (see also Hannigan 1995). Applied to the population–environment debate, if we fail to understand the bigger picture in the short term, the opportunity to act strategically and get us safely to the long term may well be lost.

BACKGROUND TO THE ISSUES

The debate over the relationship between population growth and its environmental impact has a long history (Cohen 1995), but it was not until the 1950s that it started to become a major issue on the world stage. At the time, the populations of the developing countries were just beginning to grow significantly, the result of declining infant and child mortality compounding their relatively high birth rates. So too the populations of the developed countries were growing at an increasing rate, as the baby boom (or postwar marriage boom) got under way. Population projections at the time showed world population numbers rising rapidly, with estimates for the years 2000 and 2050 around 7 billion and 12 billion respectively. Population doubling times were then in the vicinity of forty years – fifty in the developed countries and twenty-seven in the less developed – and they were decreasing (US Bureau of the Census 2004). Relatedly, concerns were being raised regarding the ability of the developing countries to feed their growing populations, and of the entire world to cope with the environmental impacts of increased agricultural production (see Carson 1962) and other economic activity writ large.

The debate gathered momentum in the 1960s and early 1970s with the publication of three significant studies: *The Population Bomb* (Ehrlich 1968), *The Tragedy of the Commons* (Hardin 1968), and *The Limits to Growth* (Meadows et al. 1974). With minor qualifications (e.g. Ehrlich & Ehrlich 1970: chs 2, 3), the neo-Malthusian 'doomsday' themes of these works told more or less the same story: under current conditions, world population numbers would soon (at least, within a century) overshoot the planet's carrying capacity, with dire consequences for humanity in the form of mass poverty, starvation, sickness and death. In short, unless appropriate interventions could be rapidly instituted (see Meadows et al. 1974: 24), life could again become, as Thomas Hobbes had proposed two centuries earlier, 'solitary, poor, nasty, brutish and short'.

These concerns were well founded. By 1971 world population growth rates had risen to 2 per cent per annum (from around 0.5 per cent at the turn of the century), implying a doubling time of thirty-five years. In the developed countries, birth rates were falling and growth rates had slowed to less than 1 per cent, with doubling times of around seventy-eight years. But in the developing countries, where 73 per cent of the world's population then lived, both had accelerated (reflecting further reductions in infant and child mortality, and an increasing momentum effect – see below), with growth rates of around 2.8 per cent per annum implying a doubling time of around twenty-five years (US Bureau of the Census 2004).

These trends generated a flurry of activity in terms of the measurement of *carrying capacity*. In its general formulation, carrying capacity is held to be 'the number of individuals who can be supported in a given land area over the long term without degrading the physical, ecological, cultural and social environment' (see Miller 1992 in Cohen 1995: 424). However, this basic concept has many limitations, most of which can be observed in the related $I = P.A.T$ model for evaluating environmental impact (Impact = Population × Affluence × Technology; see Ehrlich & Holdren 1971: 1212). Uppermost among these limitations are the lack of boundaries around the geographical areas that populations (P) actually draw upon for their resources and dump their wastes in, and interactions between the model's Affluence (A) and Technology (T) components (Hayes 1995). It is very much a 'local' model.

These limitations led to the development of the now more widely used *ecological footprint* concept, generally defined as the land and water that is required to support indefinitely the material standard of living of a given human population, using the prevailing technology (Rees 1996; Wackernagel & Rees 1996). The ecological footprint concept turns the concept of carrying capacity upside down, moving the debate from how many people a given area of land and water can support at a particular standard of living, to how much land and water – measured in terms of acres or hectares per person – a given population draw on to support themselves at a particular standard of living, *irrespective of where that land and water is.*

However, even the ecological footprint concept remains narrowly conceptualised, because it fails to engage with the longer-term implications of the different age structures *between* populations (e.g. between developed and developing countries), and changing age structures *within* populations, over time (see Figure 6.1).

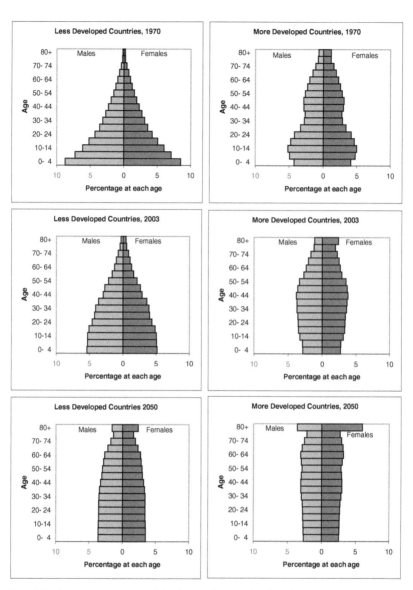

Fig. 6.1   Age–sex structures of the less and more developed countries (percentage at each age), c. 1970, 2003, 2050
Source: US Census Bureau International Data Base

In the first instance, these age structures impact on the types of resources that populations draw upon, and the types of wastes and environmental impacts they generate. In the early 1970s the median ages of the developed and developing countries were around 28 and 18 years respectively. That is, half of the populations of these regions were under or around the age at which family formation normally begins. Such youthful age structures imply heavy demand on energy-intensive resources, such as housing, vehicles, and transport. By contrast, the median ages of these regions are currently around thirty-eight and twenty-five years respectively, and are projected to increase to around forty-five and thirty-six years by 2050 (US Bureau of the Census 2004). Structurally older populations are likely to have greater proportions already housed (or seeking smaller houses) and already owning vehicles, and so on. Of course there are many other issues to be considered in terms of the relative resource needs and environmental impacts of younger and older populations, but it is precisely these factors that need to be incorporated into future population–environment modelling.

In the second instance, the 'ageing' age structures contain within them a *momentum of decline*, as the increasing numbers of elderly (and deaths) eventually come to exceed the reducing number of births. To understand this outcome we turn to a brief outline of the demographic transition.

The demographic transition refers to the well-documented decline in the birth and death rates of the now developed countries, a phenomenon that is also well under way in the developing countries. Before the onset of the transition (around 1750 and 1950 in the developed and developing countries respectively), birth and death rates were both high and essentially cancelled each other out, resulting in zero population growth. At the end of the transition, birth and death rates are low, and again cancel each other out to give zero population growth (at least theoretically). But *during* the transition, the infant death rate falls, keeping more babies alive and generating an excess of births over deaths and significant population growth. Most of this *transitional growth* occurs at the younger ages, and the age structure becomes more youthful. This phenomenon is the primary cause of the population 'explosion' described in Ehrlich's *Population Bomb*, not high birth rates per se. It is also worth noting that it was just beginning when Malthus wrote his famous work in 1798 – although he would not have realised it – and it would partly explain the empirical increase in numbers that so concerned him.

However, the population explosion also has a second cause, known as the 'momentum effect' (Keyfitz 1971). This is the growth potential contained within an age structure after the birth rate has begun to fall. Even if a population's birth rate fell immediately to the levels required for the exact replacement of each generation (2.1 births per woman), that population would generally continue to grow in size for at least one generation, because each successive cohort reaching reproductive age tends to be larger than its predecessor (this is due to the higher fertility and falling infant mortality that were extant when the reproductive age cohort itself was born). The overall outcome is that, for a short time, the combination of falling birth rates per woman and falling infant mortality produces successively larger birth cohorts, thereby compounding (adding momentum to) population growth.

Eventually, however, the momentum effect ceases as fertility continues to decline. Declining fertility and increasingly smaller reproductive age cohorts contribute fewer and fewer births to the population age structure, and cause it to contract at its base. Concomitantly, the proportion at the older ages increases, resulting in structural population ageing.

Theoretically, the reaching of replacement-level fertility was supposed to herald the end of the demographic transition. It was expected to bring with it a return to the historical situation of zero population growth, or even incipient natural decline (a *temporary* period of population decline caused by deaths outnumbering births, the outcome of increased numbers of elderly vis-à-vis falling fertility). However, in most of the more developed countries, fertility has either fallen or is continuing to fall well below replacement level, and incipient decline is likely to become population implosion.

Today, as Figure 6.2 indicates, world population growth rates are down to around 1.14 per cent per annum, implying that world population doubling time has stretched out to sixty-one years. In the developing countries the growth rate is 1.35 per cent per annum and doubling time fifty-two years, while in the developed world these are 0.28 per cent and 250 years. World population growth is likely to be down to 0.40 per cent per annum by the middle of the century, and, as theorised, zero or negative by the century's end.

KEY DEBATES

Many qualifications and sub-perspectives on the three main population–environment themes – that population growth is 'good for', 'bad for', or 'neutral to' the environment – exist. However, their main

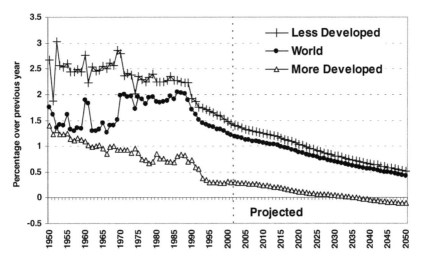

Fig. 6.2   Population growth rates (annual percentage)
Source: US Census International Data Base

premises can be gleaned from the writings of a few key theorists and commentators.

Population growth is 'bad' for the environment

The classic formulation of the 'population growth is bad for the environment' thesis – the perspective underlying the above-noted work of Ehrlich, Hardin, and Meadows and associates – has its foundations in the equally well-known work of the Reverend Thomas Malthus. Writing in 1798, Malthus' *First Essay on Population* (1996) was directed not so much at the tragedy of environmental degradation and depletion, but at human degradation. His argument was that, because of the underlying and constant 'passion between the sexes', population has the potential to grow exponentially (2-4-8-16), but, because land is finite, food and other essential resources can grow only arithmetically (1-2-3-4). The growing imbalance between the two eventually causes populations to overshoot their resources, and for many (human) lives to end in poverty, vice and misery, suffering and death, whether by direct starvation or illness, or as the result of tensions and wars brought about by competition for scarce resources. The spectre of these 'positive checks' to population growth caused Malthus to appeal for moral restraint; specifically, late marriage or permanent celibacy. These social responses he termed 'preventive checks' – practices that would limit family size and population numbers, and thereby competition for resources.

The main connection between Malthus' arguments and their place in the environmental debate relates to his belief that the natural consequence of population growth was increasing poverty. Where Adam Smith (writing in the 1770s) had argued that population growth was a *consequence* of the demand for labour, Malthus argued that population pressure preceded this demand, and that a surplus of labour would force down wages. In turn, low wages would reduce the ability of many people to marry and have children. However, low wages and an excess of workers would also mean that landowning cultivators could employ more labour, increase the area under cultivation, and thus increase the supply of food. Malthus saw this process as a cycle, following which more people would again be in a position to marry and bear children, population growth would resume, and before long population pressure would again outstrip resources. Each increase in the food supply thus meant (to Malthus) that increasing numbers would live in poverty, until their numbers were again reduced to their original levels by either the positive or preventive checks.

While Malthus' focus was the ensuing degradation of humanity, such cycles would have clear implications for the environment: each population-driven increase in the area of land under cultivation and in the technologies employed to produce and transport the food means a potential increase in environmental degradation and resource depletion. Similarly, each increase in the number of people living in poverty has direct environmental implications, as those affected are forced to survive as best they can, often using and damaging 'the commons' with little ability to care about the impact of their actions (Hardin 1968; Panayotou 1994).

These interconnections form the basis of the population–environment debate. Indeed, with one key qualification – and it is fair to say, much critique – Malthus' arguments have been and remain uppermost in the debate. The qualification concerns his rejection of contraceptive devices and practices such as abortion as the means by which to limit population numbers. For Malthus, such 'improper arts' were immoral; for his successors – the neo-Malthusians – they are the imperative. Subsequently, a long line of neo-Malthusians has called for the reduction of fertility, in particular the fertility of the developing world. Among these protagonists are the creators of many of the environmental movement's founding arguments, such as Ehrlich, Hardin, and Meadows and associates, and of other important ecological concepts such as *spaceship earth* (Fuller 1971 in Jones 2003: 225) and *future eating* (Flannery 1994).

   This qualification to Malthus' original conceptualisation of 'the problem' also reminds us of one other key attribute of neo-Malthusian
ecological perspectives on population. This is that there may be a
time-lag between a population's direct demand on resources and its
environmental impact, and further, that such relationships may be
non-linear (involving feedback loops). In its classic formulation, the
Malthusian/neo-Malthusian overshoot model is simply an inverted V,
with population size on the Y-axis, and 'time' on the X-axis. Population numbers grow to a point of overshoot and then fall, returning to
their original 'balance' with resources. While many rough estimates of
resource stocks and their rates of depletion have been carried out, it is
important to note that Malthus did not actually specify a time-frame – it
is simply often assumed to be 'short-run' (Simon 1981: 272). However,
the arbitrary nature of the concept of 'short-run' does equally little to
assist. The *Limits to Growth* work of Meadows and colleagues (1974: 24)
suggests that overshoot will occur 'within the next one hundred years',
but in ecological terms it is impossible to say whether this is a short- or
long-run period.
   The time-lag issue has many important implications, and has led to
the use and institutionalisation of what is known as 'the precautionary principle' (O'Riordan & Cameron 1994). This is the risk-related
premise that because so much about environmental damage and depletion is unknown and/or cannot actually be measured, today's populations must err on the side of caution. While strongly contested by many
in pro-economic development lobbies, the principle is now enshrined
in the Stockholm and Rio Declarations, and the Kyoto Agreement,
which is, at the time of writing, still to be ratified.

Population growth is 'good' for the environment

Directly challenging the notion that population growth is bad for the
environment is the Cornucopian or 'tech-fix' perspective, that it is only
when populations grow and begin to outstrip their resources – or at least
seriously challenge them – that they develop, or 'borrow', new or better
technologies with which to provide the required resources.
   The founding argument for this perspective is typically taken to be
Esther Boserup's work on the conditions of agricultural growth (1965,
1976). She outlines the development of agricultural and administrative
technologies as moving through increasingly sophisticated stages, from
labour-intensive *land-using* subsistence systems (such as food-gathering
and long-fallow agriculture) to organisation-intensive *land-saving*

(e.g. multi-cropping) systems. At each stage, larger populations can be supported, taking population and resources to a new balance.

However, in contrast to a number of earlier theorists who claim that populations develop or borrow new technologies in order to support their *a priori* growing numbers (e.g. Lee 1970 in Boserup 1976: 29), Boserup claimed that it is only when a population has grown and its density is high enough to support the specialisation of labour (Durkheim's social differentiation) that there can be a shift from land-using to land-saving technologies. Following this shift, there is the potential to convert periodic surpluses in food and crafts, services and technologies (such as irrigation and transport) to capital. Hence, for Boserup, there is no development or transference of technology from one population to another as long as the latter's population size permits the continued use of old technology, whereas an increase in population size brings with it the opportunity for 'economic development'.

Again the connection between these ideas and their later incorporation into the population–environment debate was not spelled out in Boserup's original work but entered the arena through the work of others. The most notable of these is Julian Simon's prominent work *The Ultimate Resource*, which began with the old adage that 'necessity is the mother of invention' (Simon 1981: 14). Directly – and throughout, provocatively – challenging neo-Malthusian calls for smaller populations, Simon reasoned that large and/or growing populations are the planet's greatest resource because they provide the possibility of, among many other things, both economic growth and environmental protection. The mechanisms by which the latter is (theoretically) achieved are economies of scale that would speed the development of substitutions for non-renewable resources. Larger populations would also contain within them more 'experts' to create the new technologies and environmentally friendly practices and policies. These propositions lie at the heart of the widely used 'tech-fix' perspective.

Importantly, the temporal and political context of Simon's publication cannot go without comment. Between 1974 and 1984 a dramatic reversal occurred in the attitude of the US Government towards the issue of population growth in the developing countries. In the 1970s, both attitudes and the US population policy program were heavily weighted towards reducing population growth by supplying contraceptives (Gulhati & Bates 1994: 49; see also McIntosh & Finkle 1985). By 1984, the US Government was taking the position that population was in fact neutral to development and environmental impact, and that the population 'problem' could be resolved by 'sound economic policies based on free markets and individual [country's] initiatives' (Gulhati & Bates 1994: 55). At the same time the United States withdrew its funding for population control (reduction).

The extent to which Simon's work may have influenced the dramatically altered US position cannot be firmly established (Teitelbaum & Winter 1985: 97–8). But this position was informed by 'revisionist economic theories' that strongly reflect Simon's arguments (Gulhati & Bates 1994: 74), especially those concerning the removal of US funding for population control.

More recently, Lomborg (2001) argued along similar lines. Since his book has been discredited (Danish Committees on Scientific Dishonesty 2002), its empirical findings are not elaborated here. However, it is important to be aware of them as they continue to be used by many of those who desire larger populations (Lowe 2003).

Population growth is 'neutral' to the environment

The 'population is neutral' perspective has its foundations in the mid-19th-century work of Marx and Engels, despite the fact that they had little to say about population dynamics – other than to severely criticise Malthus' ideas on the topic – and virtually nothing to say about the environment.

Working from the premise that it is the form of economic organisation (the infrastructure) that determines the development of societal institutions such as 'the family' and the distribution of resources (the superstructure), Marx and Engels proposed that each historical epoch has its own law of population. Although often attributed to Marx, the argument was most fully articulated after his death in Engels' famous work *The Origin of The Family, Private Property and The State* (1884). In this thesis Engels drew on the work of W.H. Morgan to illustrate how, as each form of economic organisation gave way to another, so too the form of family changed – shifting from, for example, large extended families under the feudal mode of production to smaller nucleated families under the capitalist mode of production.

The inclusion of these arguments in the population–environment debate is made through the classic Marxian precept that it is the underlying political economy (the form of economic and social organisation and its related distribution of resources) that determines the ultimate consequences of population growth. According to Marx and Engels, under capitalism this consequence is 'overpopulation', defined as there being more potential workers than can be absorbed into the economy, with the secondary consequence of poverty (and implied environmental degradation), while under socialism the consequences are state-facilitated absorption into the economy, access to a society's resources, and, *ipso facto*, environmental protection.

Contained within this argument are several of Marx and Engels' key propositions, in particular the *reserve army of labour* generated by an excess of population preceding its demand as labour, and reduced to poverty as a result of low wages – the circular outcome of the excess of workers. The argument is simply a restatement of Malthus' ideas, but with the added injunction that under one form of political economy, capitalism, that excess population will be left to fend for itself (hence Malthus' findings), while under socialism its needs will be met.

The argument led to Marx's famous dictum that 'from each according to his [sic] ability, to each according to his need'. However, this 'needs determines rights' philosophy was soundly rejected by Hardin (1977), who argued that if each is given (food and shelter) according to need – which may mean the needs of many children – there is no incentive for people to have fewer children. As a result, both family size and population are likely to be larger under socialism than capitalism, with attendant implications for environmental degradation – especially where the resources are drawn from a commons. When everyone has free access to a commons, the incentive is to use (and abuse) the resource to one's own advantage (see also Panayotou 1994 on the 'rule of capture', whereby the larger the family, the more a commons' resources can be captured and converted to private goods).

According to Hardin, the outcome of these arrangements is the opposite of Adam Smith's famous 'invisible hand', whereby individuals attending to their own needs inadvertently create a greater public good – a balance in the supply and demand of goods and services. By contrast, when drawing on a commons, the behaviour of many individuals acting in their own interests is environmentally damaging, and not at all in the interests of the collective: 'Freedom in a commons brings ruin to all' (Hardin 1977: 20). Thus, for Hardin (and Panayotou), it is the private ownership of land associated with capitalism – although not necessarily under contemporary laissez-faire arrangements – that is ultimately environmentally protective.

Largely ignoring these complexities, scholars working from the 'population is neutral' perspective return to the basic Marxian precept that the organisation of social and economic life determines a population's environmental impact. Typical of these approaches is Cohen's (1995) question of how many people *under what conditions*, although Cohen may not consider his perspective to be Marxian/Socialist. Cohen (1995: 262) lists eleven sub-questions concerning human choices, including: how many people with what average level and distribution of material well-being, with what technology, domestic and international political institutions, domestic and international demographic arrangements, physical, chemical and biological environments, values, tastes and fashions. As each social, economic, cultural, political *and* demographic constraint is factored in, a population's carrying capacity changes, typically reducing (see also Heilig 1994: 255).

Using somewhat fewer variables, similar contextual/conditional approaches are also implicit in the I=P.A.T and ecological footprint models outlined earlier, which acknowledge the role of human activities and choices in generating environmental impacts. Notably, analysts working within neo-Malthusian and Cornucopian theoretical frameworks often use such models.

Accordingly, it is worth noting that the work of Meadows and associates (1974), while typically labelled neo-Malthusian, pays central attention to the possibilities for humans to avert the global catastrophes foreshadowed in the Club of Rome's *Limits to Growth* report. Interestingly, these aspects of the report have never been given as much attention as their overall 'doomsday' message, perhaps because of their 'unrespectable' association with what can only be described as socialist propositions for greater global equality. It is thus worth quoting the report's brief conclusions in full (Meadows et al. 1974: 34):

1. If the present growth trends in world population, industrialization, pollution, food production, and resource depletion continue unchanged, the limits to growth on this planet will be reached sometime within the next one hundred years. The most probable result will be a rather sudden and uncontrollable decline in both population and industrial capacity.

2. It is possible to alter these growth trends and to establish a condition of ecological and economic stability that is sustainable far into the future. The state of global equilibrium could be designed so that the basic material needs of each person on earth are satisfied and each person had an equal opportunity to realize his [sic] individual human potential.

3. If the world's people decide to strive for this second outcome rather than the first, the sooner they begin working to attain it, the greater will be their chances of success.

FUTURE DIRECTIONS

This chapter has outlined the complex role of population in the population–environment debate, and argued for a more sophisticated approach. Currently, the three main themes that characterise the debate are based on very naive appreciations of population and its dynamics. In them, 'population' means growth in the size of a population. It does not mean the changes in age structure that have been unfolding (and will continue to unfold) as a result of the demographic transition; or that the explosive transitional growth that has occurred since the 1950s is a one-off effect of declining infant and child mortality and its related age-structural 'momentum effect'; or that this growth will soon end and in all probability will never occur again – that, in short, the population clock is about to stop ticking.

As these trends unfold, new ways of looking at the population–environment relationship will need to be developed. Before world population numbers peak and either steady off or decline, around the year 2100, lies substantial – albeit decelerating – growth. If a further 3 billion people are going to be added to present numbers, and virtually all of these are going to live in the developing regions, there is much work to be done. Current economic and social practices strongly indicate that things cannot continue as they are – everywhere there is evidence of the increased environmental degradation, depletion and poverty spoken of by the neo-Malthusians (see Hawken 1993; Hawken et al. 1999: 8). This is true even if there is the ability to reduce these effects as the Cornucopians and Marxian/Socialists argue. On the one hand, technology will be stretched to its limits to offset both the increased numbers of people and their presumably increased entry into the global economic system. On the other, the ageing of the world's populations may offer some respite, as smaller proportions of slower-growing populations will require less energy-intensive resources such as new housing and transport. But we also do not know if we have already gone beyond the limits, and may already be in overshoot (Van Dieren 1995: xi).

Just as the causes of the world's population–environment problems are complex and multidimensional, so too the answers need to be multidisciplinary and based on sound evidence. This evidence may be found in each of the debate's themes outlined above, *if* the proposed fourth dimension explicitly incorporating the demographic transition – let's call it a *transitional perspective* – is added to the scenario. Scholars may each allocate different weights to the various perspectives, depending on their own disciplinary knowledge, analytical skills and ideological

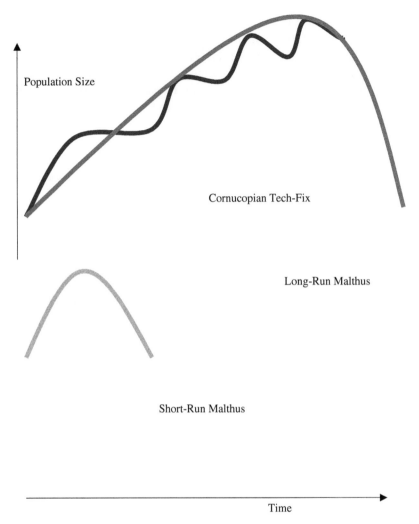

Population Size

Cornucopian Tech-Fix

Long-Run Malthus

Short-Run Malthus

Time

Fig. 6.3 Comparing short- and long-run Malthusian cycles with Cornucopian tech-fix cycles

perspectives, but it is essential that the analytical terrain becomes *all of the above*, not either/or.

The argument is illustrated in Figure 6.3. If we begin with Malthus' inverted 'V' in a short-run time-frame, we can see how populations might reach a point of overshoot and then fall back to their original balance with resources – that is, if they are dependent on local resources, as implied in I=P.A.T models. If we then superimpose on that diagram the Boserupian/Cornucopian argument – as populations grow they both borrow or develop new technology, and 'develop' economically – we can see how populations might keep reaching new balances between numbers and resources. Here we can bring to mind the idea that populations do in fact draw on global resources – the ecological footprint argument – so the time-frame *can* be longer. However, if we then similarly place Malthus' arguments in this longer and more globally resourced time-frame, we can see the possibility that he may have been right all along – that there may be a point at which the 'tech-fix' solution will no longer suffice, and population will decline due to overshoot. This could certainly occur with an additional 3 billion people to accommodate on the planet in a relatively short (say 50–100 year) time-frame, even as natural population decline waits in the wings.

The resolution to the conceptual problem is to add into this model, first the proposed 'transitional' perspective, which holds that population dynamics ultimately contain within them an internal logic of growth and then decline (but for different reasons from those proposed by Malthus), and finally the Marxian/Socialist argument that, in order to accommodate the remaining transitional growth *and* ensure an ecologically sustainable future, a more equitable planning and sharing of resources is imperative.

This, then, is the work awaiting today's students and tomorrow's scholars. Population growth is about to disappear, but not before it increases dramatically over the forthcoming half-century. Considerations of population in the population–environment debate must be urgently redirected from their long-held preoccupation with population size and growth, to population composition and natural decline. And the work must be multidisciplinary: whatever one's theoretical or ideological perspective, there is only one planet and all must work together to ensure its survival.

Discussion Questions

1. What are the three main perspectives (or themes) underlying the 'population–environment' debate?
2. How do each of these themes conceptualise the population variable?
3. What is 'transitional growth'?
4. What is the 'momentum effect'?
5. How does an age structure change from 'youthful' to 'old'?
6. In what way could Cohen's and Meadows' arguments be said to be 'socialist'?
7. What is the 'fourth pillar' proposed in this chapter?
8. Why is a 'fourth pillar' proposed?

9. In what way might the differing perspectives be considered 'political'?

10. Can you describe the argument contained in Figure 6.3?

Glossary of Terms

Birth Rate: the Total Fertility Rate (TFR) is the average number of live children a woman would have across her lifetime if she were to experience the age-specific fertility rates extant in that year, at prevailing levels of life expectancy.

Deep Greens: 'Deep Greens' (or 'Green Greens') hold the somewhat liberal or utopian view that in order to achieve ecological sustainability it is up to individuals to change the way they do things (Pepper 1993).

Demographic Convergence: the developing countries are 'catching up' with the developed countries. In the 1950s, the world's median individual lived in a country in which the birth rate was 5.6 births per woman; it is now 2.3 – rapidly approaching the generational replacement rate of 2.1 births per woman, which is expected to be reached around the end of the present decade (Wilson 2001a: 165).

Future eating: Flannery (1994) argues that human populations have always and everywhere drawn upon or destroyed ('eaten') resources needed for the sustenance of future populations.

Natural decline occurs when deaths exceed births.

Population Clock: the world population clock at http://www.census.gov/cgi-bin/ipc/popclockw gives daily updates.

Population Composition: the age, sex and ethnic composition of modern populations is typically in a state of flux, due to changes in birth, death, and migration rates. This was not the case in the (pre-demographic transition) past, where birth and death rates remained largely unchanged over the longer term, and relatively few people migrated.

Red Greens: Red Greens are Greens who see the achievement of ecological sustainability requiring government intervention, and typically hold socialist views about greater equity (Pepper 1993).

Spaceship Earth: many writers have argued that the earth is a 'materially limited and closed system' (e.g. Jones 2003: 225) and should be seen as a spacecraft, carrying one ecological system that is carrying one population.

References

Bongaarts, J., and R. Bulatao (eds) 2000, *Beyond Six Billion: Forecasting the world's population, panel on population projections.* Committee on Population, Commission on Behavioural and Social Sciences and Education, National Research Council, Washington DC: National Academy Press.
Boserup, E. 1965, *The Conditions of Agricultural Growth*, Chicago IL: Aldine.
Boserup, E. 1976, 'Environment, population and technology in primitive societies', *Population and Development Review* 1976: 21–36.
Carson, R. 1962, *Silent Spring*, London: Hamish Hamilton.

Coale, A. 1972, 'How a population ages or grows younger'. In R. Freedman (ed.) *Population: The Vital Revolution*, New York: Doubleday-Anchor, pp. 47–58.

Cohen, J. 1995, *How Many People Can the Earth Support?* New York: W.W. Norton & Co.

Danish Committees on Scientific Dishonesty 2002, Decision regarding complaints against Bjorn Lomborg. http://www.forsk.dk/uvvu/nyt/udtaldebat/bl_decision.htm (visited 14 January 2004).

Ehrlich, P. 1971/1968, *The Population Bomb*, London: Pan Books.

Ehrlich, P., and A.H. Ehrlich 1970, *Population, Resources, Environment: Issues in Human Ecology*, San Francisco CA: W.H. Freeman & Co.

Ehrlich, P., and A.H. Ehrlich 1990, *The Population Explosion*, New York: Simon & Schuster.

Ehrlich, P., and J.P. Holdren 1971, 'Impact of population growth', *Science* 171: 1212–17.

Engels, F. 1884, *The Origin of the Family, Private Property and The State, in the Light of Researches by W.H. Morgan*, transl. from 4th Russian edn (1891) by A. West and D. Torr (1931), Sydney: Current Book Distributors.

Flannery, T. 1994, *The Future Eaters: An ecological history of the Australasian lands and people*, Sydney: Reed Books.

Gulhati, K., and L.M. Bates 1994, 'Developing countries and the international population debate: politics and pragmatism'. In R. Cassen (ed.) *Population and Development: Old debates, new conclusions*, New Brunswick NJ: Transaction Publishers, pp. 47–77.

Hannigan, J.A. 1995, *Environmental Sociology: A social constructionist perspective*, London: Routledge.

Hardin, G. 1968, 'The Tragedy of the Commons', *Science*, 162: 1243–8.

Hardin, G. 1977, 'What Marx missed'. In G. Hardin and J. Baden (eds) *Managing the Commons*, San Francisco CA: W.H. Freeman & Co., pp. 3–52.

Hardin, G. 1986, 'Cultural carrying capacity: a biological approach to human problems', *BioScience* 36(9): 599–606.

Hawken, P. 1993, 'The Creation of Waste'. In *The Ecology of Commerce. A declaration of sustainability*, New York: Harper Business, pp. 37–55.

Hawken, P., A. Lovins and L.H. Lovins 1999, *Natural Capitalism: The next industrial revolution*, London: Earthscan.

Hayes, A. 1995, 'On defining the problem in population and environment'. Paper presented to the Annual Meeting of the Population Association of America, San Francisco, 6–8 April.

Heilig, G.K. 1994, 'How many people can be fed on earth?' In Lutz, *The Future Population of the World*.

Jones, A.R. 2003, 'Population goals and ecological strategies for spaceship earth', *Journal of Population Research* 20(2): 223–34.

Keyfitz, N. 1971, 'On the momentum of population growth', *Demography* 8: 71–80.

Lomborg, B. 2001, *The Skeptical Environmentalist: Measuring the real state of the world*, Cambridge University Press.

Lowe, I. 2003, 'Neither sceptical nor an environmentalist', *Earthbeat*, Radio National (Australia), Broadcast 11 October 2003. www.abc.net.au/rn/science/earth/stories/s965805.htm (accessed 4 November 2003).

Lutz, W. (ed.) 1996, *The Future Population of the World: What can we assume today?* rev. edn, London: Earthscan. www.iiasa.ac.at/Research/POP/docs/Population_Projections_Results (accessed 30 May 2001).

Lutz, W., W. Sanderson and S. Sherbov 2001, 'The end of world population growth', *Nature* 412: 543–5.

Lutz, W., S. Sherbov and W. Sanderson 2003, 'The end of population growth in Asia', *Journal of Population Research* 20(1): 125–42.

Malthus, T. (1996/1798) *An Essay on the Principles of Population as it Affects the Future Improvement of Society, with Remarks on the Speculations of Mr. Godwin, Mr. Condorcet and Other Writers*, London: Macmillan/New York: St Martin's Press.

Marx, K. (1976/1867), *Capital*, vol. 1, Harmondsworth, UK: Penguin.

Meadows, D.H., D.L. Meadows and J. Randers 1992, *Beyond the Limits: Global collapse or a sustainable future*, London: Earthscan.

Meadows, D.H., D.L. Meadows, J. Randers and W. Bherens III 1974, *The Limits to Growth: A report for the Club of Rome's project on the predicament of mankind*, 2nd edn, New York: Universe Books.

McIntosh, C., and J. Finkle 1985, 'Demographic rationalism and political systems', *International Union for the Scientific Study of Population Conference*, Florence, pp. 319–29.

McNicoll, G. 2000, 'Reflections on "replacement migration"', *People and Place* 8(4): 1–13.

O'Riordan, T., and J. Cameron (eds) 1994, *Interpreting the Precautionary Principle*, Earthscan. http://dieoff.org/page31.htm (accessed 10 October 2004).

Panayotou, T. 1994, 'Population, environment and development nexus'. In R. Cassen (ed.) *Population and Development: Old debates, new conclusions*, New Brunswick NJ: Transaction Publishers, pp. 149–80.

Pepper, D. 1993, *Eco-Socialism: From deep ecology to social justice*, London: Routledge.

Rees, W.E. 1996, 'Revisiting carrying capacity: area-based indicators of sustainability', *Population and Environment* 17: 195–215.

Simon, J. 1981, *The Ultimate Resource*, Princeton University Press.

Teitelbaum, M.S., and J.M. Winter 1985, *The Fear of Population Decline*, Orlando FL: Academic Press Inc.

Thomas, W.I., and D.S. Thomas 1928, *The Child in America: Behaviour problems and programs*, USA: Knopf.

United Nations 2000, *Replacement Migration*, UN Secretariat, Dept Economic and Social Affairs.

US Bureau of the Census (2004) International Data Base. www.census.gov/cgi-bin/ipc/idbagg (visited 12 January 2004).

Van Dieren, W. (ed.) 1995, *Taking Nature Into Account: A report to the Club of Rome*, New York: Springer-Verlag.

Wackernagel, M., and W. Rees 1996, *Our Ecological Footprint: Reducing human impact on earth*, Gabriola Island BC: New Society Publishers.

Weeks, J.R. 2002, *Population: An introduction to concepts and issues*, 8th edn, Belmont CA: Wadsworth/Thompson Learning.

Wilson, C. 2001a, 'On the scale of global demographic convergence 1950–2000', *Population and Development Review* 27(1): 155–71.

Wilson, C. 2001b, Review of J. Bongaarts and R.A. Bulatao (eds) 'Beyond six billion: forecasting the world's population (2000)', *Journal of Population Research* 18(1): 78–80.

CHAPTER SEVEN

INEQUALITY, SOCIAL DIFFERENCES AND ENVIRONMENTAL RESOURCES

Roberta Julian

What happens when 'environmentalism' meets the 'politics of differ- ence'? At the very least, an intellectually and politically exciting aspect of environmental sociology comes into view, one that focuses on issues of 'environmental justice'. The issues that become visible at such a point of intersection pose a radical challenge to mainstream under- standings of environmentalism, offer a powerful critique of contempo- rary (capitalist) social relations, and highlight the need to develop a new form of transformative politics in the 'age of globalisation'.

The postmodernist/post-structuralist turn in sociology has forced sociologists, somewhat belatedly, to take issues of 'difference' seriously in the analysis of social process. The trend here has been towards a focus on the 'politics of identity' with its insistence on the importance of listening to the voices of the 'other': ethnic and racial minorities, women, and people with disabilities. While this field of sociology exam- ines power relations, projects within this genre typically aim to uncover the underlying relations of domination and subordination through deconstruction. In doing so, they demonstrate the heterogeneous, frag- mented, and ultimately contingent nature of social life. These insights, which illuminate the flux and flow of social life, are crucial for a sound understanding of the complexities of contemporary societal conditions. There is a tendency in many of these studies, however, to avoid ground- ing analysis in the material conditions of the 'other'. Access to mate- rial resources (including environmental 'goods' and 'bads') is often not explored and for some the 'death of class' thesis (Pakulski & Waters 1996) is accepted at face value. It is precisely this issue of differential access to material (including environmental) resources that is funda- mental to the field of 'environmental justice'.

A number of social theorists, including socialist feminists such as Iris Marion Young (1990) and socialist geographers such as David Harvey (1996), have commented on the fact that the focus on 'difference' in such studies often leads to the neglect of similarities that exist across these differences. This poses a dilemma for political activists as it fragments rather than unites a potential constituency for political action (for a discussion of this issue in relation to feminist politics see Ganguly-Scrase & Julian 1999). The difficulties associated with addressing universal issues, while at the same time acknowledging the legitimacy of 'difference' (and the different social values and standpoints that accompany them), generate a unique set of tensions and debates in the environmental justice movement.

BACKGROUND TO THE ISSUES

In order to understand the field of environmental justice it is worth examining both the history of its development and its location within the broad range of environmental discourses.

History

Edwardo Lao Rhodes (2003: 5) opens his book *Environmental Justice in America* with the following statement:

> By one estimate, three out of every five African American households currently live near a hazardous-material storage area. Fines imposed on polluters by all levels of government in white communities in the 1980s were 46 percent higher than those imposed for violations in minority communities. Fines levied against site violations under the federal hazardous-waste statutes were 500 percent higher in white communities than fines in minority communities . . . Until the early 1990s, the U.S. Environmental Protection Agency had conducted no major studies on the possible uneven distribution of environmental cost or benefits across racial or income categories.

Similarly, Shank (2002: 3) points out that according to the National Black Environmental and Economic Justice Coordinating Committee, ethnic minorities in the United States are 50 per cent more likely than whites to live in communities with hazardous waste facilities.

Broadly speaking, 'environmental justice' refers to the 'fair distribution of environmental quality' (Low & Gleeson 1998: 104). As indicated by the figures quoted above, what has become evident in various places throughout the world is the existence of 'massive environmental *injustice*' (Low & Gleeson 1998: 130). It was in the recognition and articulation of this injustice in the United States that the foundations of the environmental justice movement were laid: 'The practice of the environmental justice movement has its origins in the inequalities of power and the way those inequalities have distinctive environmental consequences for the marginalised and impoverished, for those who may be freely denigrated as "others"' (Harvey 1996: 3).

While such environmental injustice has existed for a long time, Rhodes (2003: 6) notes that in the United States it is only since the mid-1980s that 'there has been a growing recognition that persons of color in both urban and rural areas may be exposed to much greater environmental risks than the American population as a whole'. This form of environmental injustice thus became known as 'environmental racism'.

Vigorous protests in the 1980s took place in the wake of some serious incidents which culminated in the First National People of Color Environmental Leadership Summit held in Washington DC in 1991 (Harvey 1996: 369; Low & Gleeson 1998: 108). This summit, attended by over 650 activists from over 300 local grassroots groups, adopted seventeen principles of environmental justice. These included:

> Environmental justice:
>
> - affirms the sacredness of Mother Earth, ecological unity and the interdependence of all species, and the right to be free from ecological destruction.
> - calls for universal protection from nuclear testing, extraction, production and disposal of toxic/hazardous wastes and poisons and nuclear testing that threaten the fundamental right to clean air, land, water and food.
> - demands the right to participate as equal partners at every level of decision making including needs assessment, planning, implementation, enforcement, and evaluation.
> - affirms the right of all workers to a safe and healthy work environment, without being forced to choose between an unsafe livelihood and unemployment. It also affirms the right of those who work at home to be free from environmental hazards.
> - affirms the need for urban and rural ecological practices to clean up and rebuild our cities and rural areas in balance with nature, honoring the cultural integrity of all our communities, and providing fair access for all to the full range of resources. (Rhodes 2003: 213–15; Grossman 1994)

The environmental justice movement continues to address the themes articulated in these principles. It is a grassroots movement (comprising, for example, 'people of colour', working-class women, children and the poor) that acknowledges the existence of different (and some-times competing) social values and standpoints among its disparate 'members'.

It is this concern with difference and the particular that constitutes both the strengths and the weaknesses of the environmental justice movement. Given the global scope, and some would argue the uni-versal concerns (Beck 1992), of contemporary environmental risk, it is not surprising to find that tensions between the particular and the uni-versal lie at the heart of the environmental justice movement. Some have argued that such tensions have generated a new environmental paradigm (Rhodes 2003) which provides the basis for new forms of political activism (Harvey 1996) as well as the impetus for institutional transformation 'in order to open global institutions to the dialectic of social justice' (Low & Gleeson 1998: 199).

Location within environmental discourses

The environmental justice movement can be described as a movement of 'resistance' in that it has arisen in opposition to the mainstream envi-ronmental movement in the United States (Hofrichter 1993). One way of demonstrating this oppositional positioning is to compare the dis-course of environmental justice with the dominant discourses on envi-ronmental issues (see Harvey 1996).

Environmental management: the 'standard' view

According to Harvey (1996), the dominant discourse on environmen-tal issues in advanced capitalist societies is one that sees environmental quality as dependent on market forces, and environmental 'problems' as outcomes that need to be 'managed'. Economic development is viewed as fundamental to human development and there is a belief that environmental concerns should not get in the way of 'progress' (in other words, capital accumulation). Environmental problems are seen as incidents to be addressed case by case 'after the event'. Typical 'solu-tions' include environmental clean-ups of particular sites rather than proactive interventions (Harvey 1996: 373–4).

The assumptions of neo-classical economics underpin this discourse. As Harvey states:

> Under the standard view, the basic rights of private property and of profit maximization are not fundamentally challenged. Concerns for environmental justice (if they exist at all) are kept strictly subservient to concerns for economic efficiency, continuous growth, and capital

accumulation . . . The only serious question is how best to manage the environment for capital accumulation, economic efficiency, and growth. (1996: 375)

This standard view has 'a substantial record of successes to its credit' (Harvey 1996: 376), particularly in public health and in contemporary efforts to improve air and water quality, but its limitations have led to the development of alternative discourses on environmental issues.

Ecological modernisation

One of these alternative discourses is that of ecological modernisation. A central premise of ecological modernisation (see Hajer 1995) is that 'economic activity systematically produces environmental harm (disruptions of "nature") and that society should therefore adopt a proactive stance with respect to environmental regulation and ecological controls' (Harvey 1996: 377).

There is a greater emphasis on the irreversibility of environmental problems than in the standard approach, and 'a rising recognition that unintended ecological consequences of human activity can be far-reaching, long-lasting, and potentially damaging' (Harvey 1996: 377). There is an associated shift to viewing environmental problems as beyond the borders, and thus the political institutions, of the nation-state. Environmental problems are viewed as global problems (acid rain, global warming, and ozone holes). The work of the German sociologist Ulrich Beck has been influential in this discourse. Importantly, however, Beck (1992) emphasises the universality of environmental risk in contemporary society.

Economic growth is still a central platform, but concern over environmental equity (distributive justice) is given greater emphasis than in the standard view (Harvey 1996: 379). In order to achieve its goal of contributing 'both to growth and global distributive justice simultaneously' (Harvey 1996: 379) there is an emphasis on the need to establish global institutional and regulatory responses (see e.g. Low & Gleeson 1998). State regulation needs to be strengthened and institutional structures need to be established at the supra-national level to address the global dimensions of environmental risk (Beck 1992).

While the rhetoric of economic modernisation has much popular appeal, and its emphasis on regulatory action may 'curb the possibilities for uncontrolled capital accumulation', Harvey warns that 'it is also a discourse that can rather too easily be corrupted into yet another discursive representation of dominant forms of economic power. It can be appropriated by multinational corporations to legitimize a global grab to manage all the world's resources' (1996: 382).

Environmental justice

Harvey (1996: 385) argues that the environmental justice discourse is 'radically at odds' with these dominant discourses. It has a distinct set of characteristics that leads Harvey to comment that it is 'far less amenable to corporate or governmental cooptation' (1996: 385) and for Rhodes (2003) to argue that it constitutes 'a new paradigm' in environmentalism. Its distinctive standpoint is apparent in the following three themes.

First, the environmental justice discourse challenges the dominant discourses by placing *inequalities* in the distribution of environmental quality 'at the top of the environmental agenda' (Harvey 1996: 385). One consequence is that its supporters are drawn from quite different social backgrounds from those in the mainstream environmental movement (Rhodes 2003). The environmental justice movement comprises a wide range of marginalised and disempowered people (the poor, people of colour, women) who are most affected by inequalities in environmental conditions. In the United States, it has included urban African-American and Latino communities and native American peoples residing on traditional lands (Low & Gleeson 1998: 107). Since environmental justice and its causes are seen differently by various groups within the environmental justice movement, this leads to 'interpretive tensions . . . across the themes of class, race, gender, and national identity' (Harvey 1996: 387; see Bullard 1993).

Second, the environmental justice discourse is critical of many mainstream environmental groups because of their 'focus on the fate of "nature" rather than humans' (Harvey 1996: 386). Scholars such as Rhodes (2003: 11) have noted that the differential impact of environmental policy and practices across racial, ethnic, income and gender 'has not been a major concern of the environmental movement'. Similarly, activists have pointed out that mainstream environmentalism has concentrated on the ecological concerns of the white middle class rather than issues such as the disproportionate burden of toxic contamination on minority communities (Low & Gleeson 1998: 107). As Taylor (1992 cited in Harvey 1996: 386) observes:

> The more established environmental organizations do fight issues of survival [but] . . . these survival debates are not linked to rural and urban poverty and quality of life issues. If it is discovered that birds have lost their nesting sites, then environmentalists go to great expense and

lengths to erect nesting boxes and find alternative breeding sites for them . . . but we have yet to see an environmental group champion the cause of homelessness in humans or joblessness as issues on which it will spend vast resources. It is a strange paradox that a movement which . . . worries about the continued survival of nature (particularly loss of habitat problems), somehow forgets about the survival of humans (especially those who have lost their 'habitats' and 'food sources').

Third, the environmental justice discourse is sceptical of rational arguments from experts and professionals about environmental impacts. These experts have often been co-opted by those in power to 'either deny, question, or diminish what were known or strongly felt to be serious health effects deriving from unequal exposure' to environmental hazards (Harvey 1996: 386). Thus the environmental justice discourse searches for 'an alternative rationality' (Harvey 1996: 386). In doing so, it shifts attention away from questions of 'what is legally, scientifically, and pragmatically possible?' to 'what is morally correct?' (Harvey 1996: 389).

KEY DEBATES

Given its oppositional location vis-à-vis dominant environmental discourses and the mainstream environmental movement, it is not surprising to find that the field of environmental justice is rife with tension and debate. One area of debate revolves around the meaning of the term 'environmental justice'. Other questions include: should the movement focus on 'difference' and the particular concerns of specific communities or should it develop a more universal approach to addressing matters of environmental concern; in what ways are the local and global dimensions of environmental justice/injustice related; and is the focus on prevention or 'cure'? A consideration of all these questions leads to the positioning of environmental justice at the nexus between environmental and social issues. Debates around these questions provide the dynamic momentum for the environmental justice movement.

Defining environmental justice

The meaning of the term 'environmental justice' is both complex and contested. In the first instance, it can be distinguished from 'ecological justice' by its focus on human populations (and the relationships between sub-populations) rather than the relationship between humans and nature. In the United States, environmental justice has predominantly referred to 'fairness in the distribution of environmental well-being' (Low & Gleeson 1998: 102). Rhodes (2003: 19) offers the following definition:

> The fair treatment of all races, cultures, incomes, and educational levels with respect to the development, implementation, and enforcement of environmental laws, regulations, and policies. Fair treatment implies that no populations of people should be forced to shoulder a disproportionate share of the negative environmental impacts of pollution or environmental hazards, or be denied a proportionate share of the positive benefits of environmental regulation or programs, due to lack of political or economic strength.

This definition emphasises that environmental *in*justice exists when there is an unequal distribution of 'environmental quality' across differing populations. It also acknowledges that this unequal distribution is a political issue: differential access to power lies at the heart of environmental injustice.

Since the 1970s, studies have examined the relationship between residential location and proximity to desirable or undesirable land-uses and public services that impact on well-being. Many of these studies identified race as a key factor affecting the distribution of hazardous wastes (United Church of Christ 1987; Bullard 1990, 1992; Bullard & Wright 1990; Bryant & Mohai 1992; Mohai & Bryant 1992; Adeola 1994). Low & Gleeson (1998: 104) conclude that 'a collusion between markets and racially discriminatory anti-ecological local politics has produced a racialized pattern of risk in the United States, meaning that many urban colored communities now bear a disproportionate share of the environmental risks that arise from the nation's hazardous industries'.

Traditionally, environmental justice has focused on the *distribution* of environmental quality. But environmental justice can be examined from two different vantage points: distribution and production. The emphasis to date has been on 'environmental equity' with its focus on 'the equitable distribution of negative externalities' (Low & Gleeson 1998: 112). Critics of the 'environmental equity' approach (Heiman 1996; Lake 1996), however, stress the need to analyse the structural *sources* of injustice and to avoid getting bogged down in 'the quicksand of distributional politics' (Low & Gleeson 1998: 113). They stress the need to shift attention away from issues of distribution (after the fact) to the processes of production. Furthermore, they argue that minorities must be included in decision-making in the sphere of production if environmental justice is to be achieved. As Heiman states:

'Environmental justice demands more than mere exposure equity . . . it must incorporate democratic participation in the production decision itself' (1996 cited in Low & Gleeson 1998: 112–13).

Tensions within environmental justice

This emphasis on difference, power relations, patterns of risk, social values and empowerment gives rise to a number of key tensions within the environmental justice movement.

Particularism versus universalism

The tension between the particular and the universal is one of these. Postmodernist critiques have led to the questioning of 'universals' as part and parcel of the critique of the Enlightenment project more broadly. The idea that there is some universal agreement on the meaning of justice (and thus environmental justice) is therefore anathema to postmodernists. They draw attention to the fact that 'there can be no universal conception of justice to which we can appeal . . . There are only particular, competing, fragmented, and heterogeneous conceptions of and discourses about justice which arise out of the particular situations of those involved' (Harvey 1996: 342).

This emphasis on 'difference' and the importance of grounding conceptions of justice in the particular are fundamental to an understanding of the power relations that are inherent in environmental conditions and regulations. These insights are central to the environmental justice movement.

Nevertheless, the focus on 'difference' also has potential deficiencies. First, it leads to a tendency to reify the particular and to elevate the standpoints of specific populations (such as people of colour, women, ethnic groups, the poor and people with a disability) over any considerations that may be more universal. The environmental justice movement is thus known for its 'militant particularism' in that its focus is on the particular struggles of specific communities. It is primarily concerned with access to environmental resources among people whose livelihoods are threatened and therefore tends to engage in place-bound politics (see e.g. Strangio 2001).

Another pitfall that may arise from a recognition of divergent individual and communal values is the tendency to adopt a position of cultural relativism with respect to environmental problems. As Low & Gleeson (1998: 194) note: 'There are *objective* dangers arising from contemporary industrialism, in the form of toxic wastes and other hazards, which cannot be socially distributed merely through a system of culturally derived preferences . . . Too often cultural relativism is a mask for anti-democratic politics and even localised tyranny.'

The way in which the economic management model attempts to deal with difference demonstrates the problems that may arise here. In this model, communities are assumed to have differing social values that lead them to make different choices regarding land-use. This includes choices with respect to LULUs (locally unwanted land-uses) including not only industrial activities, but also residential land-uses for 'socially undesirable' people, such as homes for ex-prisoners, deinstitutionalised mental patients, and people with AIDS. Communities that 'choose' to host LULUs are awarded financial compensation for bearing the burden of environmental hazards. Consumer preferences are taken into account in a cost–benefit analysis.

However, this approach does not take into account the existence of power inequalities and the consequent constraints on 'choice'. These are clearly evident in studies that have demonstrated the pressures placed on Native American communities to accept toxic landfills on their land in return for financial compensation (Bullard 1993). Similarly, Cutter (1995) reports on a situation in which 'the Apache nation sought the establishment of a private nuclear waste facility on its own territory in New Mexico in return for monetary compensation' (Low & Gleeson 1998: 118). Harvey (1996) and Low & Gleeson (1998) have both emphasised that such financial compensation will be of immediate appeal to leaders of communities suffering from economic insecurity.

Thus it is important that the recognition of 'difference' takes place in conjunction with the recognition of underlying power relations. Otherwise this 'monetisation of risk' simply allows structural inequalities to be exploited by risk producers (see O'Hare et al. 1983; Armour 1991; Boerner & Lambert 1995). The outcomes are 'environmental blackmail' and 'localised tyranny' disguised as cultural tolerance. Nothing is done to change the structural conditions that give rise to this exploitation. At the same time, since compensation is typically a one-off event and does not compensate succeeding inhabitants of communities that accept LULUs, then it may 'entrench intergenerational inequity' (Low & Gleeson 1998: 117).

Furthermore, this utilitarian 'solution' to the LULU problem presupposes perfect access to information. In reality, however, differing communities have varying access to scientific knowledge and other relevant information (Low & Gleeson 1998: 117–18). In broad terms, the problem with this utilitarian view is that it includes 'no appreciation of social power, and the asymmetries which derive from class, race and gender difference' (Low & Gleeson 1998: 118).

Central to environmental justice is the recognition of such asymmetrical power relations. This leads to debates within the movement over the 'causes' of environmental and social injustice. Some (e.g. Harvey 1996) argue that they are fundamentally class-based, some (e.g. Bullard 1994) argue that it is race that is the key dimension producing differential outcomes, while others (e.g. Rhodes 2003) suggest that environmental inequity is the result of a multifaceted process so that engaging in debates over the primary cause is a non-productive exercise.

Given the emphasis on diversity within the movement, Harvey (1996) has noted that if the concerns of environmental justice are to be taken up by those in power, the movement needs to appeal to some more universal values. The problem here, however, is that it is difficult to determine 'universal prescriptions of what is a fair distribution of environmental quality' (Low & Gleeson 1998: 104).

Local versus global

Another key tension in environmental justice relates to the scale of analysis. This creative tension is reflected in a critique of political processes at the level of the nation-state and an emphasis on the need to address issues of environmental justice at the global level.

Of particular concern to the environmental justice movement are 'the health risks posed by hazardous industries and waste management activities' (Low & Gleeson, 1998: 114). This has translated into a concern with the problem of where to site LULUs. As awareness of the health hazards of LULUs has increased, so has popular antipathy for residential proximity to LULUs. The acronym 'NIMBY' (Not In My Backyard) has been coined to describe this.

At the national level, it is the relationship between states, markets and local communities that determines distributions of environmental quality (Low & Gleeson 1998: 104). One approach adopted by the state is to establish regulatory mechanisms that, informed by scientific advice and involving community consultations, are charged with identifying the 'safest' locations for LULUs. NIMBY opposition is thus viewed as threatening the 'orderly control of industrial wastes' as well as contributing to a decline in jobs and overall economic well-being. An alternative approach involves the attempt to develop 'fair-share planning regulations which aim to establish geographic uniformity in the distribution of risk' (Low & Gleeson 1998: 116). Uniformity in outcomes is unlikely, however, given political resistance from capital and

NIMBY opposition from 'environmentally privileged classes' (Low & Gleeson 1998: 116).

Importantly, the successes of many of the grassroots campaigns in the United States have created further problems for the environmental justice movement. The worsening of racial and class disparities may be an unintended consequence of these successes (Low & Gleeson 1998: 113). As Goldman (1996 cited in Low & Gleeson 1998: 113) has noted: 'As more communities try to block sites and prevent pollution in their backyards, those with the least political and economic power will be left with an even greater share of the toxic residues from our modern society.'

Such an outcome further increases the 'otherness' of marginal communities. LULUs become concentrated in particular residential communities who are increasingly 'contaminated' with the unpleasant and risky side-effects of these land-uses (Low & Gleeson 1998: 114). Following Douglas (1978), the danger associated with such impurities further stigmatises the members of these communities and reinforces their 'otherness' vis-à-vis non-polluted communities. Images of pollution and impurity are then used as claims and counter-claims to status.

NIMBY opposition has further developed into NIABY (Not In Anyone's Backyard) and NOPE (Not On Planet Earth). One of the consequences of this rising opposition in the developed world has been that 'corporations are even more likely to move the most noxious plants to less developed countries, where even poorer communities of color will be the hosts' (Goldman 1996 cited in Low & Gleeson 1998: 119). As Pulido has noted: 'the political successes of the environmental movement in developed countries may actually accelerate the relocation of hazardous industries to developing nations. Environmental regulations are increasingly cited by US firms for their flight to more "business friendly" countries, such as Mexico' (1996 cited in Low & Gleeson 1998: 122). The result is that people in developing nations are increasingly exposed to environmental risk in both the production process and in the toxic wastes disposed of in waterways, landfills, sewerage and drainage systems and in the countryside (Low & Gleeson 1998: 123).

The situation at the national level is thus mirrored at the global level where developing countries play the same role as the poorer communities within the developed nations. This has led to what Low & Gleeson (1998: 121) refer to as the 'traffic in risk':

> The structural difference of economic needs and government regulation between the developed and developing worlds, and the absence of any supra-national body to ensure consistency in environmental

standards, has encouraged western industrial capital to shift unpopular and increasingly illegal hazard-producing activities and wastes across national boundaries to states which often define, and welcome, these transfers as 'investment'.

The uneven landscape of environmental risk thus becomes an intrinsic part of the 'uneven development' on which capitalism depends (see e.g. Friedrichs & Friedrichs 2002).

This outcome at the global level has provided the impetus for exploring new forms of transnational political organisation and new institutional frameworks for environmental regulation. Critics have recognised that 'the global political system, composed of competing nation states, may be unfitted for the task of guaranteeing environmental justice' (Low & Gleeson 1998: 120). For Low & Gleeson (1998: 131) this calls for the need to establish an administrative state at the international level to regulate the distribution of risk. At the same time, they stress that 'the political critique within developed countries must be shifted from the *spatial allocation* of risk to the *production* of risk'.

Prevention versus cure

All these tensions culminate in a concern over the goals of the environmental justice movement as a whole. Should the movement be focused on 'cure' or 'prevention'?

When the central concerns are those of environmental equity and the spatial allocation of risk then the main game is 'distributive justice'. But the limits to this game are evident as one moves from the local to the global level. Clearly, given the existing institutional and political structures of advanced capitalism, engaging in the game of 'distributional politics' tends to lead to the displacement of problems. The emphasis is on finding a 'cure' for the side-effects of environmental risk rather than addressing the underlying causes of environmental injustice.

In contrast, the environmental justice discourse highlights the limitations of any approach to environmental problems that takes the current political and institutional arrangements for granted. It recognises 'the tendency of capitalism to distribute socio-economic and environmental resources unevenly' (Low & Gleeson 1998: 106) and thus provides the impetus for a radical transformation of environmental politics. As Harvey (1996: 368) has noted, a focus on distributive justice rarely, if ever, leads to a questioning of how and why hazardous wastes are produced in the first place. For Commoner (1990) and others, 'the question of *prevention* surely should take precedence over disposal and cure of any side-effects' (Harvey 1996: 368). The posing of such questions, however, 'requires a discursive shift to the far more politically charged terrain of critique of the general characteristics of the mode of production and consumption in which we live' (Harvey 1996: 368).

This is a shift that the environmental justice movement is prepared to take. It is the reason it remains an oppositional movement and why its discourse is less likely than other environmental discourses to be appropriated by those in power.

The environmental–social nexus

Adding to this 'radical' view of environmental politics is the fact that within the rubric of environmental justice, the 'environment' itself is redefined to include a range of social issues that impact on overall well-being. Thus its agenda becomes broader while remaining grounded in the particular experiences of its members. As Harvey (1996: 399) points out: 'The principles of justice it enunciates are embedded in a particular experiential world and environmental objectives are coupled with a struggle for recognition, respect, empowerment.'

Given the diversity of positions within the environmental justice movement, it is not surprising to find that there are quite different measures of injustice within it. Krauss (1994: 270) makes this clear when he observes:

> Women's protests have different beginning places, and their analyses of environmental justice are mediated by issues of class and race. For white blue-collar women, the critique of the corporate state and the realization of a more genuine democracy are central to a vision of environmental justice . . . For women of color, it is the link between race and environment, rather than between class and environment, that characterizes definitions of environmental justice. African American women's narratives strongly link environmental justice to other social justice concerns, such as jobs, housing and crime. Environmental justice comes to mean the need to resolve the broad social inequities of race. For Native American women, environmental justice is bound up with the sovereignty of the indigenous peoples. (cited in Harvey 1996: 387)

The outcome is twofold. On the one hand, it leads to debates and tensions within the movement across the dimensions of class, race, gender and national identity. On the other hand, the environment itself is redefined to include 'the totality of life conditions in our communities – air and water, safe jobs for all at decent wages, housing, education, health care, humane prisons, equity, justice' (Southern Organizing

Committee for Economic and Social Justice 1992 cited in Harvey 1996: 391).

The environmental justice movement connects environmental justice and social justice in unique ways. In doing so, it comes to focus much of its attention on the cleaning up and rebuilding of urban environments. This concern with the urban further distinguishes environmental justice from mainstream environmentalism, which focuses on the non-urban, especially the wilderness.

FUTURE DIRECTIONS

The environmental justice discourse is little known outside the United States. However, given the increased political concern with environmental risk in the last decade, and the increased significance of environmental issues within the discipline of sociology, this is likely to change.

The tensions identified and discussed in this chapter are central to politics and sociology. How can environmental justice move beyond a concern with 'distributive justice' to address the *production* of environmental risk? How can it develop an approach that is grounded in local experiences but appeals to general principles that would provide the momentum for political activism at a global level?

One way of resolving these tensions, proffered by Harvey (1996) and others (e.g. Young 1990), is to interpret universality to mean 'the participation and inclusion of everyone in moral and social life' rather than the 'adoption of a general point of view that leaves behind particular affiliations, feelings, commitments, and desires' (Young 1990: 105). Similarly, for Heiman (1996), the environmental justice ideal centres on 'community empowerment and access to the resources necessary for an active role in decisions affecting people's lives' (cited in Low & Gleeson 1998: 113).

From this perspective, environmental justice is as much about process as it is about outcomes. For Low & Gleeson, justice is 'an open-ended dialectical process' that contributes to discursive democracy. They advocate a shift from competitive politics that dominates the world today, to 'deliberative politics, in which the search for the truth of the human condition is sought'. However, a prerequisite for this shift is 'the relief of the competitive insecurity into which the peoples of the world have been thrown' (Low & Gleeson 1998: 196, 198).

Discursive democracy relies on the principle that people involved in decision-making are under no coercion at all. In Habermas' terms, the absence of any kind of force is necessary for the process of 'reaching understanding' (Low & Gleeson 1998: 202), an essential component of communicative action:

> While Habermas never seeks to distill universal principles, he does insist on a process . . . of free and unfettered communication occurring in the public sphere of civil society. His aim is to democratize communicative action to the point where it can be the bearer of powerful ethical principles, such as those pertaining to justice. (Harvey 1996: 353)

Thus universality cannot and should not be avoided. Rather, 'universality must be construed in dialectical relation with particularity' (Harvey 1996: 362).

Athanasiou (1996) highlights the crucial importance of deliberative politics when he states: 'We inhabit a paradox. Our age is tragic, and catastrophe does threaten, but though the future is obscure, it does not come to us inexorable and inescapable. Our tragedy lies in the richness of the available alternatives, and in the fact that so few of them are ever seriously explored' (1996: 306–7).

In conclusion, as noted in the introduction to this chapter, when 'environmentalism' meets the 'politics of difference' an exciting aspect of environmental sociology becomes visible. The focus to date, however, has been on race and class-based minorities. There is much research to be done with respect to minorities based on gender, sexuality and disability. Furthermore, outside the United States, the sociology of environmental justice is in its infancy. Despite the urgency of its political and social agenda, it awaits the focused attention of sociologists worldwide.

Discussion Questions

1. What are the key features of 'environmental justice'?
2. How does the discourse of environmental justice differ from environmental management, ecological modernisation, and the 'wise use' discourses?
3. In what ways do the concerns of the environmental justice movement differ from the concerns of mainstream environmental organisations?
4. What are the strengths of the ecological modernisation discourse? What are its limitations?
5. Provide some examples of environmental injustice that have occurred within the nation-state (identify which nation-state).
6. Provide some examples of environmental injustice that have occurred at the global level.
7. To what extent are transnational corporations and/or governments responsible for the uneven distribution of environmental risk at the global level?

8. Brainstorm some strategies that could be used by the environmental justice movement to provide a bridge between local and global environmental concerns.

9. Would the development of a regulatory institution at the supra-national level solve the problems of environmental injustice?

10. 'Exploring alternative modes of production, consumption and distribution as well as alternative modes of environmental transformation is a waste of time given current economic and political conditions.' Discuss.

Glossary of Terms

Communicative action: an ideal form of free and unfettered communication occurring in the public sphere of social life.

Discursive democracy: an ideal form of democracy based on uncoerced discourse and perfect information.

Ecological modernisation: an approach to environmental issues that focuses on the production of environmental harm.

Environmental justice: a radical discourse that addresses issues of inequality in environmental conditions, policies and outcomes at national and global levels.

Environmentalism: a broad set of discourses that focuses on issues relating to the environment.

Particularism: a concern with the localised and specific cultural values and practices of a community or population.

The 'politics of difference': the contemporary politics that has emerged around issues arising from a recognition of the cultural diversity that exists in advanced modernity. Among others, these politics focus on issues of race, ethnicity, gender, sexuality, and/or disability.

Postmodernism: an intellectual and aesthetic movement that focuses on explorations of 'difference' between cultures, institutions and individuals.

Social justice: a contested term in philosophy and social science that refers to 'good' and 'proper' relationships between ourselves and others.

Universalism: a concern with the commonalities and generalities that can unite people across differences such as those based on class, race, ethnicity and gender.

References

Adeola, E.O. 1994, 'Environmental hazards, health and racial inequity in hazardous waste distribution', *Environment and Behavior* 26(1): 99–126.

Armour, A. 1991, 'The siting of locally unwanted land uses: towards a cooperative approach', *Progress in Planning* 35(1): 1–74.

Athanasiou, T. 1996, *Divided Planet: The ecology of rich and poor*, Boston MA: Little, Brown & Co.

Beck, U. 1992, *Risk Society*, London: Sage.

Boerner, C., and T. Lambert 1995, 'Environmental injustice', *Public Interest* Winter: 61–82.

Bryant, B., and P. Mohai (eds) 1992, *Race and the Incidence of Environmental Hazards*, Boulder CO: Westview Press.

Bullard, R. 1990, *Dumping in Dixie*, Boulder CO: Westview Press.

Bullard, R. 1992, 'Environmental blackmail in minority communities'. In Bryant and Mohai, *Race and the Incidence of Environmental Hazards*, pp. 82–95.

Bullard, R. (ed.) 1993, *Confronting Environmental Racism: Voices from the grassroots*, Boston MA: Southend Press.

Bullard, R. (ed.) 1994, *Unequal Protection: Environmental justice and communities of color*, San Francisco CA: Sierra Club Books.

Bullard, R., and B.H. Wright 1990, 'Toxic waste and the African American community', *Urban League Review* 13(1–2): 67–75.

Commoner, B. 1990, *Making Peace with the Planet*, New York: Pantheon Books.

Cutter, S. 1995, 'Race, class and environmental justice', *Progress in Human Geography* 19(1): 111–22.

Douglas, M. 1978, *Purity and Danger*, London: Routledge & Kegan Paul.

Friedrichs, D.O., and J. Friedrichs 2002, 'The World Bank and Crimes of Globalization: A Case Study', *Social Justice* 29(1–2): 13–36.

Ganguly-Scrase, R., and R. Julian 1999, 'Minority women and the experiences of migration', *Women's Studies International Forum* 21(6): 633–48

Grossman, K. 1994, 'The people of color environmental summit'. In Bullard, *Unequal Protection*.

Hajer, M. 1995, *The Politics of Environmental Discourse: Ecological modernization and the policy process*, Oxford: Oxford University Press.

Harvey, D. 1996, *Justice, Nature and the Geography of Difference*, Oxford: Blackwell.

Heiman, M.K. 1996, 'Race, waste, and class: new perspectives on environmental justice', *Antipode* 28(2): 111–21.

Hofrichter, R. 1993, 'Introduction'. In R. Hofrichter and L. Gibbs (eds) *Toxic Struggles: The Theory and Practice of Environmental Justice*, Philadelphia PA: New Society Publishers, pp. 1–10.

Lake, R.W. 1996, 'Volunteers, NIMBYs, and environmental justice: dilemmas of democratic practice', *Antipode* 28(2): 161–74.

Low, N., and B. Gleeson 1998, *Justice, Society and Nature: An exploration of political ecology*, London: Routledge.

Mohai, P., and B. Bryant 1992, 'Environmental injustice: weighing race and class as factors in the distribution of environmental hazards', *University of Colorado Law Review* 63: 921–32.

O'Hare, M., L. Bacow and D. Sanderson 1983, *Facility Siting and Public Opposition*, New York: Van Norstrand Reinhold.

Pakulski, J., and M. Waters 1996, *The Death of Class*, London: Sage.

Rhodes, E.L. 2003, *Environmental Justice in America*, Bloomington IN: Indiana University Press.

Shank, G. 2002, 'Overview: globalization and environmental harm', *Social Justice* 29(102): 1–12.

Strangio, P. 2001, *No Toxic Dump! A triumph for grassroots democracy and environmental justice*, Sydney: Pluto.

United Church of Christ 1987, *Toxic Wastes and Race in the United States*, New York: United Church of Christ Commission for Racial Justice.

Young, I.M. 1990, *Justice and the Politics of Difference*, Princeton University Press.

CHAPTER EIGHT

SUSTAINABLE TECHNOLOGY: BEYOND FIX AND FIXATION

Aidan Davison

Today's world looks like no other that has gone before. Cars, televisions, air-conditioners, computers: objects unimaginable only a few generations ago, yet now so familiar as to make the idea of life without them almost as unimaginable. The simplest of activities depend on the unseen genius of hydro-electric turbines, cables overhead, underground and undersea, combine harvesters, satellites – all knit together by global flows of electronic information. This is a world populated by transgenic crops, cloned animals and humans conceived in test tubes. A world in which the already ill-fitting categories 'natural' and 'artificial' will be of ever less help, where replacements for worn-out body parts may soon be grown from our tissue, and where surgeons and soldiers alike benefit from ever more powerful tools.

Perhaps the most remarkable thing about this world, however, is not the objects that fill it so much as the fact that change itself seems to be its unifying constant. The surfaces around us are altered ceaselessly in the race to keep up to date. Time seems to accelerate and space to shrink, as the basis of social order becomes mobility itself (Urry 2000). The proliferation of mobile telephones in little more than a decade, combined with their continual innovation and capacity for making fashion statements, tells a story typical of this world's capacity for social change, not to mention economic growth.

A more disturbing story is to be found in news that the number of people who are overweight has grown to be more than 1.1 billion, matching the numbers of those who live in hunger (Gardner & Halweil 2000). In the United States alone, 61 per cent of the adult population is overweight, with around 40 per cent of this group classed as obese (Gardner 2001). To interpret this story we need to remember that although the second half of the 20th century saw the world economy grow more than 700 per cent (Worldwatch Institute 2002: 59), the gap between the wealth of the richest 20 per cent and the poorest 20 per cent of the human population increased two and a half times during this period, reaching a ratio of 74 to 1 by 1997 (UNEP 2002: 35). More to the point, if the entire population of nearly 6.5 billion were to live as this wealthy 20 per cent does – in affluent nations such as Australia, Britain, France, Germany, Japan and the United States, and within elites in even the poorest nations – it would require the natural resources of somewhere between four and eight planets like our own (UNEP 2002: 36). It is not hard to imagine that in such a world conflicts over access to natural resources and distribution of environmental risks will increase in frequency and intensity.

In part, these facts remind us of the colonial history on which many modern technological triumphs are built, burdening the present with a legacy of ecological damage, social inequality and cultural imperialism. Aeroplanes and the Internet may bring the globe within reach to most of us with access to university education, but these technologies remain inaccessible to a majority of the human population (UNEP 2002: 36–7). While we all face ecological problems of global scale, the minority living in abundance do so in circumstances very different from those of the majority living in poverty. The facts of malnutrition are also disturbing, however, because they remind us that affluent modern life does not equate with social, physiological or psychological well-being in any straightforward way, for abundance, too, has its dangers.

The question of what sustainability can mean in a world where powerful currents of technological change are defining new patterns of excess and scarcity at bewildering speed is a commanding one. The literature addressing this question has grown vast since the early 1970s (Dobson 1999; Harris et al. 2001). Arising jointly out of concern about ecological damage done to a finite planet and about inequalities in modern forms of social development, this literature now takes as its common purpose the integration of environmental, economic, social and cultural objectives into a seamless and stable platform for public policy (Dale 2001; Kohn et al. 1999). There is, however, wide and passionate disagreement about the nature of these objectives. The resultant debate is complex, involving many different contesting positions, and no one topic has succeeded in producing conflict and misunderstanding as thoroughly as that of technological progress.

This chapter reviews controversy over the role of technology in the creation of more sustainable societies. The discussion begins by introducing the two explanations of technology, and the themes of fix and fixation to which they give rise, that have dominated these debates. These explanations, *technologies as neutral servants* and *technologies as autonomous masters*, are shown to be the source of endless conflict. Apparently opposed, both explanations ignore the many ways in which technological means and human ends interact. In response, the final section introduces a third explanation, that of *technology as social practice*. In exploring the prospects for sustainable technological futures, this explanation invites inquiry into technology as both a product and a producer of social order and cultural meaning.

BACKGROUND TO THE ISSUES

Since 1980, news of environmental crisis has produced a great deal of discussion about the need for technological management of natural environments, much of it under the heading of policy for sustainable development. But it has also provoked widely held and acutely felt fears and hopes about technology (Davison 2001: 11–89; Prugh et al. 2001). This is not surprising. As we shall see, although modern scientific and economic institutions routinely assume technology to be a value-neutral tool, utopian dreams and dystopian nightmares about technology have long had powerful influence in the Western imagination (Winner 1977; Noble 1997). On one side gather prophets of 'simplicity' who see in modern technology a self-destructive addiction, a deadly fixation laying waste to people and planet with equal abandon (e.g. Mander 1991; Mills 1997; McKibben 2003). On the other side gather prophets of a 'Golden Age' who see in technological progress a way of fixing, permanently, the problems of society and ecology, thereby building a safe future with few if any limits (e.g. Kelly 1994; Easterbrook 1995; Lomborg 2001). Labels such as 'technocrat' and 'Luddite' have become barbed weapons with which to attack opponents in a debate whose heat invites combative metaphors. It seems we are impelled to declare our allegiance: are we for technology or against it? Is technology a force for good or a force for evil?

In these conditions the demonising and the canonising of technology is common. Caricatures of naked hippies returning to 'the cave' or films about 'brave new technocrats' doing the bidding of computers are familiar subjects of popular culture (Falzon 2002: 149–80). Careful visions of the future, those capable of reminding us of enduring possibilities in today's political choices, are shouldered aside by the easy fatalism of apocalyptic or complacent claims that the script of the future is already written (Prugh et al. 2001: 21–46).

Social scientists have contributed to the conflict such all-or-nothing positions create, positions demanding of almost religious faith, to the extent that they have opted out of these value-laden debates. Seeking the safe ground of objectivity, many have joined natural scientists in treating technologies simply as physical facts. It has often been assumed, therefore, that rational conversation about sustainable technology is best led by engineers and economists, and in a language closer to mathematics than to politics or ethics (e.g. Weaver et al. 2000).

It is also true, however, that the presumption that social life is best understood through the lens of scientific objectivity has been challenged in recent decades. With this has come growing interest in the complexity and ambivalence of phenomena wrapped up in the phrase 'modern technology'. A multidisciplinary coalition of perspectives now recognises the dangers of trying to pull issues of fact and technology away from those of value and politics. Since the early 1970s this inquiry has spread rapidly under a range of titles, the most common of which is perhaps that of science, technology and society studies (STS). Within this field, philosophical analyses have largely inquired into the human essence of technology (Borgmann 1984; Winner 1986; Higgs et al. 2000). In contrast, political, sociological and historical analyses have explored the social construction of specific technologies in particular contexts, emphasising issues of gender, race, class, identity, culture and ecology (Cockburn & Dilic 1994; Jasanoff et al. 1995; Reynolds & Cutcliffe 1997). While 'essentialist' and 'constructivist' approaches have often developed independently of each other, they have begun to converge into a rich conversation that does not simply bring together different disciplines but challenges the validity of the boundaries that define them in the first place (Feenberg 1999; Cutcliffe & Mitcham 2001; Ihde & Selinger 2003).

Drawing from STS literature, the following section investigates two dominant and apparently antithetical sets of assumptions about technology, to be discussed here under the headings of *instrumentalism* and *determinism*. These assumptions have been applied to growing concern about environmental issues to produce proscriptions for quick tech-fixes and warnings about deadly fixations. Tracing the sources of this conflict, we consider how to move beyond it and towards approaches to sustainability that do not separate technology from social contests over political ideals, moral values and cultural worldviews.

KEY DEBATES

Instrumentalism

Instrumentalism names the claim that technology is simply the sum of all those tools, those artefacts, which humans use to advance their interests in life. This view is most often presented as commonsensical – after all, guns don't prowl the streets by themselves – and is the explanation embedded in modern scientific and economic institutions. It draws on two distinct but convergent sources of justification that we can call *naturalist* and *rationalist*. Naturalist justifications see technology as a fact of nature, and thus socially neutral, changing according to laws akin to those of natural evolution. Rationalist justifications, in contrast, see technology as a human and not a natural phenomenon, albeit one still socially neutral, the product of objective rationality. While the former has been important in entrenching instrumentalism in modern life, the latter is explicit in the formal definition of technology as 'the scientific study of the practical or industrial arts' (Oxford English Dictionary 1989). This definition supports the perception that technologies are empty conduits for ideas. Technology is, in this view, applied science and, seeming to lack any meaning of its own, has 'come to mean everything and anything; it therefore threatens to mean nothing' (Winner 1977: 10).

The instrumentalist representation of technologies as unquestionably loyal servants dominates sustainable development policy. Consider the claim in the sustainable development manifesto – the Brundtland Commission's *Our Common Future* – that with 'careful management new and emerging technologies offer enormous opportunities for raising productivity and living standards, for improving health, and for conserving the natural resource base' (WCED 1987: 217). Assuming that sustainability hinges on objective management of technologies, rather than on any properties inherent in technologies themselves, the Commission placed its faith in the continued evolution of presently dominant technological systems: 'Information technology . . . can help improve the productivity, energy and resource efficiency, and organizational structure of industry . . . The products of genetic engineering could dramatically improve human and animal health . . . Advances in space technology . . . also hold promise for the Third World' (WCED 1987: 217–18).

The World Resources Institute stripped this faith back to its core in 1991, asserting that 'technological change has contributed most to the expansion of wealth and productivity. Properly channelled, it could hold the key to environmental sustainability as well' (Heaton et al. 1991: vii, ix). The following year, technological efficiency was the biggest hope shining through *Agenda 21*, the United Nations' action plan for sustainable development, despite the fact that the term 'technology' itself was notable mostly for its absence (Davison 2001: 25).

Indeed, *Agenda 21* is a powerful example of the paradox of instrumentalism: namely, that the more technology becomes an organising principle of social policy, the less there seems to be to say about it other than how to achieve it.

Confidence in the pursuit of the efficient maximisation of production as the shortest path to sustainable development was elaborated during the 1990s, chiefly through the idea of eco-efficiency advocated by the World Business Council for Sustainable Development (Schmidheiny 1992; de Simone & Popoff 1997), but also, in a more subtle way, under the heading of 'ecological modernisation theory' (Mol & Sonnenfeld 2000). The decade closed with Paul Hawken, Amory Lovins and Hunter Lovins' (1999) blueprint for 'Natural Capitalism', based on a 'Factor 10' increase in resource use efficiency, although only two years earlier the latter two authors, writing with another colleague (von Weizsäcker et al. 1997), had judged 'Factor 4' to be sufficient to realise sustainability.

Unlike the 1970s and 1980s, when environmental discussion focused on the earth's limits to growth, the entry into a new century has been marked by confidence that the only limits that matter are those imposed by the current state of technology and furthermore that sustainable development will ensure these continue to be pushed back (Davison 2001: 13–17). Earlier emphasis on strengthening and extending governmental regulation is giving way to an expanded role for responsible corporations. Armed with triple bottom line accounting, cradle-to-cradle management, closed-loop production and other eco-techniques (see Holliday et al. 2002), corporations champion the goal of profitable environmental stewardship. Unlike the three earlier UN conferences on the environment (in 1972, 1982 and 1992), the 2002 World Summit on Sustainable Development's *Plan of Implementation* assumes that sustainable consumption requires above all else increases in the efficiency of product management and that it requires little if any political and cultural negotiation about modern lifestyles, or about the global systems of production, information and finance on which they rest (UN 2002: 13–20).

Beyond instrumentalism

Instrumentalist explanations offer some insight into technology, and the agenda of eco-efficiency is capable of improving some measures of environmental quality. They are, however, unable to address many of the social causes of unsustainability for they arise out of the fallacy of taking a partial truth to be the whole truth about technology.

In maintaining the separateness of human ends and technological means, such explanations are unable to expose the values and assumptions that inform modern understandings of development, progress and sustainability.

Certainly technologies function, in part, as tools. Descriptions of these functions provide what Carl Mitcham (1994: 160) usefully calls *first-order* definitions of technology. Such definitions have clear merit. It is still the case, for instance, that television sets come with on/off switches, as well as with an increasing choice of content and modes of delivery, not to mention the fact that consumers are free not to buy these tools or to discard them. Eco-efficiency offers the promise that soon television sets – of earthier tones and textures, no doubt – will be made out of recycled materials, using fewer resources and producing less waste of lower toxicity, that they will be powered by 'cleaner' energy sources and that they will advertise an ever-growing range of 'green' products and provide interactive sustainability training via DVD.

These outcomes are to be welcomed in a world in which television is a major preoccupation. Nonetheless, this is not the whole picture! Television has also produced profound changes in human experience, giving rise to a wide set of social, or *second-order*, meanings. By becoming central to the flows of information, money, images, stories and desires in social life, television has also become central to the contest for political power and cultural legitimacy (Kubey & Csikszentmihalyi 1990). It has enabled the acceleration and expansion of the cultural practices of consumption, in the process changing the character of these practices. To take an example mentioned earlier, television is implicated in the problem of over-nutrition through the combination of physical inactivity and advertising of processed and fast foods – to which could perhaps be added, passive intellectual and emotional habits – associated with it. The social world of Australian children born in the 1980s is not comparable with that of the 1930s, in no small part because of the changes television has made possible to everyday practices. And, of course, changes arising from this innovation are inextricably enmeshed with the countless effects rippling outwards from other areas of innovation through networks of social practice. These worlds are not just physically different. They are different in the kinds of human experiences, needs and capacities they make more or less familiar.

Modern scientific and economic institutions are largely concerned with first-order or instrumental meanings of technological sustainability, imagining an impermeable wall between the objectivity of science and the messy complexity of wider society. In contrast, environmental social movements have had much to say about second-order implications of technological change, and in so doing have relied heavily on the second dominant mode of modern explanation of technology, that of technological determinism.

Determinism

Technological determinism refers to two groups of perspectives sharing two assumptions: first, that technological development is at least partly autonomous, unfolding according to its own internal forces outside the sphere of social control; second, that technological autonomy sets limits on human autonomy, thus exercising social control in its own right. Unlike instrumentalists who explain technology as a 'law' of natural evolution, determinists represent technology not as value-neutral but as political power independent of human action, entrenching some social interests and values while undermining others, thus altering the balance and direction of social development.

Put in such bald terms, determinist explanations seem unconvincing. It appears illogical to claim that the things that humans make can become masters over their makers. Why, then, have such explanations been a significant theme in modern political thought and popular culture (Winner 1977)? The simplest answer to this complex question is that technological determinism has more to do with everyday experience than it does with rational argument. It arises out of the confusing and conflicting feelings that have been provoked by technological change since the first axe made both kindling and revenge more readily available. It is instructive to read, for instance, even if it is hard to comprehend in our era, Plautus' reaction to the high-tech of his day, some twenty-three centuries ago: 'Who in this place set up a sun-dial,/To cut and hack my days so wretchedly/Into small portions' (cited in Boorstin 1983: 28).

Often discussed under the labels of *technophobia* – fear of technology – and *technophilia* – adoration of technology – strong emotional responses are integral to our embodied interaction with the technologies around us (Thayer 1995). Cars may appear as collections of objects on a factory line, but once they have become a paraplegic's means of mobility or a setting for horrific injury or a code of social status or an adolescent's rite of passage or a threat to the atmosphere they are inseparable from the conflicting interests at the centre of social life. Wrapped up as they are in this drama, it is not so difficult to believe that technologies have at least some life of their own and many inhabit our experience as friends and enemies rather than as tools.

Technological determinism ranges along a spectrum from pessimistic to optimistic versions of technological destiny. In optimistic accounts our technological dictators are benevolent. Technology becomes, in effect, the principle of evolution embodied in our species. In this vein, the environmentalist Buckminster-Fuller claimed that 'the universe is a comprehensive system of technology' (1970: 178). More recently Kelly has taken up this theme: 'As we improve our machines they will become more organic, more biological, more like life, because life is the best technology for living' (1994: 212). Easterbrook emphasises the flip side of this argument, announcing that nature 'needs us – perhaps, needs us desperately' to overcome its limitations and errors (1995: 668). This is the way, utopian determinists argue, that humanity will design out 'the age-old failures of war, poverty, hunger, debt, nationalism, and unnecessary human suffering' (Fresco & Meadows 2002: 35). Such views are not new. They can be, in part, traced back to reinterpretations of Christian belief in the European Middle Ages that were later cemented into the foundations of modern science, most notably through the writings of Francis Bacon (Noble 1999).

At the other end of the determinist spectrum the prospect of technological malevolence has been recorded famously in Mary Shelley's *Frankenstein*, Aldous Huxley's *Brave New World* and George Orwell's *Nineteen-Eighty Four*. The following observation is typical:

> The machine has got to be accepted, but it is probably better to accept it rather as one accepts a drug – that is, grudgingly and suspiciously. Like a drug, the machine is useful, dangerous and habit-forming . . . You only have to look about you at this moment to realise with what sinister speed the machine is getting us into its power. (Orwell 1936: 189)

Orwell fails to come close, however, to the antipathy D.H. Lawrence had packed, a few years earlier, into his poem 'Death is Not Evil, Evil is Mechanical', with its shattering conclusion that those seeking immortality through technology 'begin to spin round on the hub of the obscene ego/a grey void thing that goes without wandering/a machine that in itself is nothing/a centre of the evil world-soul' (1986: 248).

Such visions of technological excess owed much to earlier Romantic traditions in art and literature (Postman 1999). They were to be deepened by the technologised horror of the Second World War and the acceleration of industrial innovation it catalysed. From the late 1940s through to the 1960s, intellectuals as diverse as philosopher Martin Heidegger, theologian Jacques Ellul, historian Lewis Mumford, and critical theorist Herbert Marcuse converged on Lawrence's theme that humanity was itself in danger of becoming a machine (Davison 2001: 96–100). Inevitably, the recognition of global ecological damage that began to spread rapidly during the 1970s became a further and powerful motive in casting technology as an inhuman and unnatural force bent on destruction of life itself. At least, it did so until the goal of eco-efficiency emerged in the late 1980s as a new centre of gravity in another phase of instrumentalist optimism in technological progress.

Beyond determinism

Despite their very different conclusions about technology, instrumentalist and determinist explanations rest on a shared foundation. They both accept the binary or dualistic logic of modern rationality, given famous expression by René Descartes four centuries ago, that represents body and mind as entirely separate categories of existence, with 'the body' belonging to the inferior realm of nature and 'the mind' to the supreme realm of culture (Plumwood 1993). Viewed through the lens of dualism, technology and human, means and ends appear discontinuous. Technology is located outside the human essence, becoming either servants of ideas and morals, in the case of instrumentalism, or, in the case of determinism, inhuman forces acting on ideas and morals.

The section below explores debates over the sustainability of technological futures. Yet rather than understand technology as tools for fixing complex social problems with engineering solutions or as an addictive fixation that has humanity in the grip of a suicidal dependency, it draws on recent theoretical interest in technology as social practice. It presents technology as a key ingredient of the social conditions of identity and relationship into which ideas about sustainability must be translated.

FUTURE DIRECTIONS

An absolute, rigid distinction between culture and nature has been a central feature of modern Western traditions (Latour 1993), and one that has shaped much thinking about sustainability. Embedded within this distinction is an assumption that both culture and nature can be reduced to a single and very different universal essence. During the history of industrialisation, the technological digestion of the earth's 'resources' – the universal essence of Nature – and their reconstitution in the forms of 'civilisation' – the universal essence of Culture – defined human progress. The pendulum has now well and truly swung on this history. Expressions of nature's essential value and meaning are everywhere visible in everything from environmental philosophies to real estate spiel (Cronon 1996). Whether thought of as machinery doing a poor job and needing technological updating, or as virginal wilderness needing protection from technology, however, nature remains often understood as opposed to culture.

The conceptual separation of nature and culture is most often translated in practical terms into the view that technology is opposed to ecology as if it were a separate category of reality, the boundary between them policed by notions of 'naturalness' and 'artificiality'. As will be appreciated from the foregoing discussion, this translation is not straightforward because modern traditions have also imagined an impermeable wall between the facts of technology and the values of society. Thus while ecology and technology are thought to be fundamentally different, both have been excluded from understandings of what makes us human. Many environmentalists have sought to reclaim ecology as a source of human meaning. Yet in celebrating nature's inherent values, much environmentalism has only strengthened the perception of technology as inhuman and unnatural (Davison 2001: 63–89). Radical ecology movements, in particular, often promote a form of direct human reunion with nature seen as avoiding altogether the alienating mediation of technology (e.g. Sessions 1995). What is often lost in this yearning for an enchanted nature is recognition that all human practices, those that cherish and nurture life as much as those that seek to control and destroy it, are inherently technological.

Representations of humanity, ecology and technology as separate forms of reality lie beneath the surface of current debates about sustainability and destabilise many policy attempts to pursue integrated forms of social development. Sustainable development has become a mechanism for fitting together technological, ecological, and socio-cultural objectives, rather than questioning how these objectives became disintegrated in the first place and how this disintegration might be avoided.

This state of affairs is profoundly ironic, for at the level of everyday practice the boundaries between humanity, ecology and technology are ever more permeable. Recent developments in gene technology, for instance, provide a vivid example of the ability of contemporary technology to make 'thoroughly ambiguous the difference between natural and artificial, mind and body, self-developing and externally designed, and many other distinctions that used to apply to organisms and machines' (Haraway 1991: 152). Controversy over gene technologies now threatens to burn out of control as these practices move out of laboratories into farms, factories and hospitals, touted as the solution to everything from human disease, starvation and depression to

biodiversity, pollution and waste management (Tokar 2001). Instrumentalist assurances that gene technologies are just another set of tools
needing rational management hold little weight in the face of powerful
resistance focused on at least six key issues:

1. the immorality and arrogance of seeking to redesign and commodify
   life;

2. the global risks to human and ecological health from a reductionist
   technology seeking control over complex living systems by manipulating what is thought to be their basic building blocks;

3. the inadequacy of capitalist motives of profit, competition and consumer preference in providing equitable human benefit and in tack-
   ling basic rather than trivial human needs;

4. the legal control of gene technology, including 'genetic information', by a relatively small number of large transnational corporations;

5. the capacity of gene technology to further widen the gap between
   wealthy and poor;

6. the use of gene technology for violent social purposes, and in the
   service of political authoritarianism in general.

These concerns are vitally important and demanding of serious political
debate and action (Hindmarsh & Lawrence 2001; Bridge et al. 2003).
Such debate and action are confused, however, by determinist representations of technology which suggest that the core threat posed by
gene technology is that to an essential human (e.g. Fukuyama 2002) or
natural (e.g. McKibben 2003) purity, to their sanctity, and which seek
to redraw a line in the sand between the organic and the technological.
   Alternatives such as 'organic' forms of agriculture or 'holistic'
medicine may well be more sustainable than many emerging forms of
gene technology. But they are not less technological, or more natural, in any essential sense. They represent crucially different forms of
biotechnological social practice that need to be articulated as alternative and positive visions of technology – visions affirming that empa-
thy and interconnection are as much technological possibilities as are
control and alienation. Many forms of modern technology have done
great damage to ecological and social systems by assuming they can be
controlled as if humans stood outside them. The environmentalist celebration of the human location deep within these systems does not,
however, mean that technology ought to be demonised and rejected.
Rather, it demands inquiry into the ways technologies make possible

human embeddedness within these systems. Understanding technology as the means of belonging within networks of sustaining relationships ensures that technical issues of efficiency and control are unavoidably joined to social issues of sufficiency and moral purpose.

Insight into technology as social practice reveals more than just the ways in which technologies are developed and used in the context of social beliefs, values and goals. It asks how technologies are also constitutive of such beliefs, values and goals in the first place. Put simply, it asks: how are we being built as we build our world? Most importantly, such insight does not just provide greater powers of description. Inquiry into technology as social practice empowers groups and individuals to take political and moral responsibility for the building of human possibilities.

Objectifications of technology give rise to the strange result that the habitats humans build are not understood to be inherently part of their humanity (Davison 2004). Human agents inhabit technological space almost as strangers, asking of the objects around them: are you tool or tyrant? This question is of only limited use in taking practical responsibility for the challenge of sustainability. Technologies of genetics, biology, energy, matter and information cannot be neatly sorted into good and bad, or sustainable and unsustainable, piles. Produced within militaristic – or unjust or colonising or wasteful or racist or patriarchal, etc. – social practices, renewable energy technologies, sustainable forms of agriculture and other 'green' techniques may reduce some forms of ecological risk, but they may also help to prop up, to sustain, an unsustaining social whole. Then again, given that some powerful voices champion nuclear power as a renewable energy source and genetic engineering as a cornerstone of sustainable agriculture (Holliday et al. 2002), such approaches are by no means certain to reduce ecological risks either.

Technologies are integral to the political and moral processes that shape social life. Furthermore, they now play a more central role in the creation of human possibilities, and in the creation of ecological and other risks, than ever before. Bruno Latour, a leading sociologist of technology, encourages us to observe how the modern project of dominating nature has created a paradoxical social reality in which mastery and predictability of any kind are ever less likely:

> Behind the tired repetition of the theme of the neutrality of 'technologies-that-are-neither-good-nor-bad-but-will-be-what-man-makes-of-them', or the theme, identical in its foundation, of

'technology-that-becomes-crazy-because-it-has-become-autonomous-and-no-longer-has-any-other-end-except-its-goalless-development', hides the fear of discovering this reality so new to modern man who has acquired the habit to dominate: there are *no masters anymore* – not even crazed technologies. (Latour 2002: 255)

This realisation offers very different understandings of sustainability from those produced by instrumentalist and determinist representations of the future as either controllable through rational planning or prefabricated through the trajectories of the present. It emphasises that technologies build the habitats through which societies continuously remake human experiences, capacities and needs in ways inherently experimental and unpredictable; in ways never fully knowable. 'If you want to keep your intentions straight, your plans inflexible, your programme of action rigid, then do not pass through any form of technological life' is Latour's wry advice (2002: 252), for technologies are the means of achieving predetermined ends, as well as the means of creating new ends.

There are many ways in which understandings of technology as experimentation in human possibilities can inform ideas of sustainability. Recent interest in social theory of risk, mobility, hybridity, networks and contested natures (e.g. Adam et al. 2000; Urry 2000; Castree & Braun 2001; Macnaghten & Urry 2001; Ihde & Selinger 2003), in particular, has much to offer discussion of technology and sustainability. Such approaches reveal questions of sustainable technology to encompass much more than the imperative of efficiency, for they reach back to the most basic concerns of human meaning. They enable us to see technological futures afresh by exposing the limits of representations of technology as tools or tyrants and by reclaiming technological choices as political and moral negotiations about the human character of social practice.

Discussion Questions

1. Give three examples of technology you would like to see in the everyday life of the 22nd century. How does each example embody your vision of 'social progress'?
2. Discuss the claim that 'genetically modified foods represent the best hope for overcoming global food insecurity'.
3. Outline one argument in favour and one against the cloning of extinct and endangered species as a means of biodiversity protection.
4. Are the phrases 'sustainable technology' and 'sustaining technology' interchangeable? Why?

5. Assess the recyclable, renewably powered, non-polluting private automobile as an example of sustainable technology.
6. Describe three representations of technology in recent popular culture in relation to the themes of instrumentalism and determinism.
7. What ought sociological study of the idea of 'eco-efficiency' encompass?
8. What effect is implementation of renewable energy technologies likely to have on the gap between the richest 20 per cent and the poorest 20 per cent of the global population during the next twenty years?
9. Explain your understanding of wilderness. How is this understanding related to your everyday experience of 'built' environments?
10. Evaluate the following proposition: 'the traditional technologies of indigenous cultures were just as advanced as those of today and can help provide solutions to global environmental problems.'

Glossary of Terms

Determinism (technological) represents technological change as autonomous, outside human control, and either socially good (utopian) or evil (dystopian).

Eco-efficiency: measure of ecological impacts and economic prosperity per unit of production.

Instrumentalism represents technology as a collection of neutral tools, or physical facts, lacking any inherent meanings, purposes, politics or values.

Luddite: deriving from the resistance of 19th-century English craft workers to industrialism, now often used pejoratively about people seen to be anti-technology.

STS: interdisciplinary field of science, technology and society studies.

Sustainability: a wide arena of debate, rather than a specific concept, asking basic questions about the relationship of: humanity to nature; present to future generations; wealthy to poor; technology to social progress; and global to local politics.

Sustainable development: narrow interpretations of sustainability focused on increased production and resource use efficiency.

Technocrat: those in positions of social authority who reduce social problems to matters of instrumental calculation and technological efficiency.

Technology: inherent dimension of human experience having both first-order (invention, production and tool use) and second-order (worldviews, politics, morals) meanings.

References

Adam, B., U. Beck and J. van Loon (eds) 2000, *The Risk Society and Beyond: Critical issues for social theory*, London/Thousand Oaks CA/New Delhi: Sage.

Boorstin, D. 1983, *The Discoverers: A history of man's search to know his world and himself*, London: J.M. Dent & Sons.

Borgmann, A. 1984, *Technology and the Character of Contemporary Life: A philosophical inquiry*, Chicago University Press.

Bridge, G., P. McManus and T. Marsden 2003, 'The next new thing? Biotechnology and its discontents', *Geoforum* 34: 165–74.

Buckminster-Fuller, R. 1970, 'Technology and the human environment'. In R. Disch (ed.) *The Ecological Conscience*, Englewood Cliffs NJ: Prentice Hall, pp. 174–80.

Castree, N., and B. Braun (eds) 2001, *Social Nature: Theory, practice and politics*, Oxford/Malden MA: Blackwell.

Cockburn, C., and R.F. Dilic 1994, *Bringing Technology Home: Gender and technology in a changing Europe*, Milton Keynes, UK/Philadelphia PA: Open University Press.

Cronon, W. (ed.) 1996, *Uncommon Ground: Rethinking the human place in nature*, New York: W.W. Norton & Co.

Cutcliffe, S., and C. Mitcham 2001, *Visions of STS: Counterpoints in science, technology and society studies*, Albany NY: State University of New York Press.

Dale, A. 2001, *At the Edge: Sustainable development in the 21st century*, Vancouver/Toronto: UBC Press.

Davison, A. 2001, *Technology and the Contested Meanings of Sustainability*, Albany NY: State University of New York Press.

Davison, A. 2004, 'Reinhabiting technology: means in ends and the practice of place', *Technology in Society* 26(1): 84–97.

de Simone, L.D., and F. Popoff 1997, *Eco-efficiency: The business link to sustainable development*, Cambridge MA/London: MIT Press.

Dobson, A. (ed.) 1999, *Fairness and Futurity: Essays on environmental sustainability and social justice*, Oxford/New York: Oxford University Press.

Easterbrook, G. 1995, *A Moment on the Earth: The coming age of environmental optimism*, New York: Viking.

Falzon, C. 2002, *Philosophy Goes to the Movies: An introduction to philosophy*, London/New York: Routledge.

Feenberg, A. 1999, *Questioning Technology*, London/New York: Routledge.

Fresco, J., and R. Meadows 2002, 'Engineering a new vision of tomorrow', *The Futurist* (Jan.–Feb.): 33–6.

Fukuyama, F. 2002, *Our Posthuman Future: Consequences of the biotechnology revolution*, New York: Farrar, Straus & Giroux.

Gardner, G. 2001, 'Being overweight now epidemic'. In Worldwatch Institute, *Vital Signs 2001*, pp. 136–7.

Gardner, G., and B. Halweil 2000, 'Overfed and underfed: the global epidemic of malnutrition'. In J.A. Peterson (ed.) *Worldwatch Paper 150*, Washington DC: Worldwatch Institute.

Haraway, D. 1991, *Simians, Cyborgs, and Women: The reinvention of nature*, New York: Routledge.

Harris, J., T. Wise, K. Gallagher and N. Goodwin (eds) 2001, *A Survey of Sustainable Development: Social and economic dimensions*, Washington DC/Covelo CA: Island Press.

Hawken, P., A.B. Lovins and L.H. Lovins 1999, *Natural Capitalism: The next industrial revolution*, London: Earthscan.

Heaton, G., R. Repetto and R. Sobin 1991, *Transforming Technology: An agenda for environmentally sustainable growth in the 21st century*, Washington DC: World Resources Institute.

Higgs, E., A. Light and D. Strong (eds) 2000, *Technology and the Good Life?* Chicago/London: University of Chicago Press.

Hindmarsh, R., and G. Lawrence (eds) 2001, *Altered Genes II: The future?* Melbourne: Scribe.

Holliday, C., S. Schmidheiny and P. Watts 2002, *Walking the Talk: The business case for sustainable development*, San Francisco CA: Berrett-Koehler.

Ihde, D., and E. Selinger (eds) 2003, *Chasing Technoscience: Matrix for materiality*. Bloomington/Indianapolis IN: Indiana University Press.

Jasanoff, S., G.E. Markle, J.C. Peterson and T. Pinch (eds) 1995, *Handbook of Science and Technology Studies*, Thousand Oaks CA/London/New Delhi: Sage.

Kelly, K. 1994, *Out of Control: The new biology of machines*, London: Fourth Estate.

Kohn, J., J. Gowdy, F. Hinterberger and J. van der Straaten (eds) 1999, *Sustainability in Question: The search for a conceptual framework*, Cheltenham, UK: Edward Elgar.

Kubey, R., and M. Csikszentmihalyi 1990, *Television and the Quality of Life: How viewing shapes everyday experience*, Hillsdale NJ: Erlbaum Associates.

Latour, B. 1993, *We Have Never Been Modern*, transl. C. Porter, Cambridge MA: Harvard University Press.

Latour, B. 2002, 'Morality and technology: the end of the means', *Theory, Culture & Society* 19(5/6): 247–60.

Lawrence, D.H. 1986, *D.H. Lawrence: Poems*, selected and introduced by Keith Sagar, rev. edn, Harmondsworth, UK: Penguin.

Lomborg, B. 2001, *The Skeptical Environmentalist: Measuring the real state of the world*, Cambridge University Press.

Macnaghten, P., & J. Urry (eds) 2001, *Bodies of Nature*, London/Thousand Oaks CA/New Delhi: Sage.

Mander, J. 1991, *In the Absence of the Sacred: The failure of technology and the survival of the Indian nations*, San Francisco CA: Sierra Club Books.

McKibben, B. 2003, *Enough: Staying human in an engineered age*, New York: Times Books.

Mills, S. (ed.) 1997, *Turning Away from Technology: A new vision for the 21st century*, San Francisco CA: Sierra Club Books.

Mitcham, C. 1994, *Thinking Through Technology: The path between engineering and philosophy*, University of Chicago Press.

Mol, A., and D. Sonnenfeld 2000, 'Ecological modernisation around the world: an introduction', *Environmental Politics* 9(1): 3–14.

Noble, D., 1999/1997, *The Religion of Technology: The divinity of man and the spirit of invention*, Harmondsworth, UK: Penguin.

Orwell, G. 1936, *The Road to Wigan Pier: The complete works of George Orwell*, vol. 5, London: Secker & Warburg.

Oxford English Dictionary (1989), Oxford: Clarendon Press.

Plumwood, V. 1993, *Feminism and the Mastery of Nature*, London/New York: Routledge.

Postman, N. 1999, *Building a Bridge to the Eighteenth Century: How the past can improve our future*, Melbourne: Scribe Publications.

Prugh, T., R. Costanza and H. Daly 2001, *The Local Politics of Global Sustainability*, Washington DC/Covelo CA: Island Press.

Reynolds, T.S., and S. Cutcliffe 1997, *Technology and the West: A historical anthology from Technology and Culture*, Chicago IL/London: University of Chicago Press.

Schmidheiny, S. 1992, *Changing Course: A global business perspective on environment and development*, Cambridge MA: MIT Press.

Sessions, G. (ed.) 1995, *Deep Ecology for the 21st Century: Readings on the philosophy and practice of the new environmentalism*, Boston MA: Shambhala.

Thayer, R. 1995, *Grey World, Green Heart: Technology, Nature, and the sustainable landscape*, Series in Sustainable Design, New York: John Wiley & Sons.

Tokar, B. (ed.) 2001, *Redesigning Life? The worldwide challenge to genetic engineering*, Melbourne: Scribe Publications.

United Nations 2002, 'Plan of implementation'. In *Report of the World Summit on Sustainable Development*. United Nations Document A/CONF.199/20* (reissued), available at www.johannesburgsummit.org (accessed 12 December 2003).

UNEP (United Nations Environment Program) 2002, *Global Environmental Outlook 3: Past, present and future perspectives*, London: Earthscan/UNEP.

Urry, J. 2000, *Sociology Beyond Societies: Mobilities for the twenty-first century*, London/New York: Routledge.

von Weizsäcker, E., A.B. Lovins and H.L. Lovins 1997, *Factor Four: Doubling wealth – halving resource use. A new report to the Club of Rome*, Sydney: Allen & Unwin.

WCED (World Commission on Environment and Development) 1987, *Our Common Future*, Oxford: Oxford University Press.

Weaver, P., L. Jansen, G. van Gootveld, E. van Spiegel and P. Vergragt 2000, *Sustainable Technology Development*, Sheffield, UK: Greenleaf Publishing.

Winner, L. 1977, *Autonomous Technology: Technics out-of-control as a theme in political thought*, Cambridge MA: MIT Press.

Winner, L. 1986, *The Whale and the Reactor: A search for limits in an age of high technology*, Chicago IL/London: University of Chicago Press.

Worldwatch Institute (ed.) 2002, *Vital Signs 2002: The trends that are shaping our future*, New York/London: W.W. Norton & Co.

CHAPTER NINE

THINK GLOBAL, ACT LOCAL

Scalar Challenges to Sustainable Development
of Marine Environments

Elaine Stratford

Among the social sciences, long-standing debates continue about the effects of economic globalisation. Part of that discussion is about the principles and practices by which to be modern *and* exhibit stewardship over economic, social and environmental well-being. Often 'sustainability' describes the principles of such care, and 'sustainable development' denotes the practices by which these are enacted.

Debating the worth and consequences of economic globalisation involves asking to what extent the state is best placed to address the challenges of modern life. This question informs concerns about democracy and citizenship. Supra-statists advocate investing more power in structures and processes of global governance and government, suggesting more centralising and authoritarian strategies for 'the greater good'. Sub-statists are equally committed to devolving power to subnational systems of decision-making that privilege the local (Wapner 1995). Advocates of both positions attribute to existing state systems the vast majority of environmental woes, calling for the reorganisation of political life and the transference of power *up* or *down* spatial scales.

Differences between supra-statists and sub-statists bring into sharp relief questions about the *scale* at which sustainable development is best deployed. The catch-cry *think global, act local* captures this uncertainty, suggesting that economic, social and environmental problems are transboundary, and the need to *engage and empower* via democratic and civic rights and responsibilities for sustainable development. Nowhere are such issues better etched than in relation to archetypal trans-boundary domains, the global commons, including the marine environment.

Over 70 per cent of Earth is aquatic: oceans, coasts and islands are gravely at risk from processes of modernisation and economic globalisation. Despite the proliferation of mechanisms to advance sustainable development at different scales, the evidence suggests that few goals of this agenda have been fulfilled.

This relative failure needs to be understood and addressed. As one response, in this chapter I explore how management regimes for marine environments are characterised by an 'implementation deficit' (Crowley 1999) which is partly attributable to tensions across scales of governance. The implementation deficit is shorthand for the failure of formal strategies of sustainable development to 'put the brakes' on environmental degradation, social dislocation and economic instability. While the deficit might be applied to various cases, in what follows I refer to marine environments, which allow an exploration of scale and some ways to understand environmental controversies.

BACKGROUND TO THE ISSUES

Marine environments

Marine environments inspire deep affective responses in people. They are crucial to global life-support systems and ecosystem services across the planet, their contribution to global climate regulation being especially important (McConnell 2002; Huber et al. 2003). Their biological and geological diversity spans coral reefs and estuaries, mangroves and wetlands, sea mounts, ocean trenches and other benthic domains (Summerhayes et al. 2002). Over a billion souls depend for their main sources of protein on seafood. Hundreds of millions rely on artisanal and commercial mariculture for their livelihoods (Cole 2003; Eagle & Barton 2003; Future Harvest 2003). Untold numbers are also directly and indirectly involved in illegal, unregulated and unreported fishing (Fallon & Stratford 2003). Coasts will continue to be most profoundly affected by such activities (Lindeboom 2002; Summerhayes et al. 2002). Small islands are often viewed as particularly vulnerable to climate change and sea-level rise (Mitchell & Hinds 1999; Pelling & Uitto 2001; Joost et al. 2002).

Marine environments are the lifespaces of seagrasses, algae, phytoplankton and other marine flora. Some, such as red seaweeds, may be useful in the treatment of pandemics such as HIV/AIDS (Burges et al. in preparation; Global Campaign for Microbicides 2003; Population Council 2003; Women's Global Health Imperative 2003). This aquatic realm is also the wellspring of rare minerals, oil and natural gas, and evidence suggests the existence in the oceans of assorted energy sources already earmarked for exploitation (Costanza 1999; Halfara & Fujitab 2002; Wells et al. 2002; Jones & Morgan 2003). Indeed, managing

deep-sea mining may be especially taxing for international environmental policy communities as both private and public interests seek
to maximise the monetary and strategic flow-on effects of such activity
(Huber et al. 2003; Smith 2003).

Among other things, marine environments are affected by military
activities (Pirtle 2000); trade and the movement of sometimes very hazardous wastes (Vanderzwaag 2002); tourism (Trist 1999; Ghina 2003);
and various categories of displaced persons (Pallis 2002). Significant
numbers of vessels are unregistered or under flags of convenience. Their
regulation is extremely difficult. Many are single-hulled and poorly
maintained. Unknown numbers illegally transport contraband and are
used for the illegal harvest of marine resources (Kullenberg 2002).

Thus marine environments are increasingly at risk (Huber et al.
2003; Smith 2003). As the risk grows so does the number of strategies of
global governance to address it. Many schemes position sustainability as
a dominant ethical guideline (or way of being) and sustainable development as a set of normative practices (or ways of doing). Despite all the
activity, there 'is growing concern that *we* are not proving as successful
as might be wished in protecting *our* planet and sustaining *our* future'
(Summerhayes et al. 2002: 1–2; emphasis added). This assessment
underscores two further dilemmas: a tendency to value the earth instru-
mentally – for what it offers people rather than for what it intrinsically
is; and a propensity to ignore how humans (and non-humans) perceive,
use or value things at different scales. Indeed, to forget scale is to over-
look crucial dimensions of engagement and empowerment – namely,
being in and *nourishing* place. Here, the term *being* does not simply signify
material presence in a landscape, but a sense of committed attachment
which, in the case of the commons, is vital in developing capacities to
nourish that which is simultaneously 'ours' and 'not ours'.

Scale and levels of governance

Scale is an elusive concept. It may refer to relative magnitude. In certain parts of the world, for example, scales of local and regional degradation of marine environments are greater than in others. Scale can
signify an ordered standard, such as might exist if improvements to
coastal management capacities shift from point x to point y in a range of
outcome measures for state-of-the-environment reporting. It can indicate a ratio between an object and a representation of that object,
as in the case of a seascape and an oceanographic map of it. Scale is
implied in design, measurement, calculation, regulation or production.
It is implied in the relativities between things: that challenge is serious,

this one is trivial; that impact was then or will be later, this impact is now; that feature is distant, this one is near.

Scale is as much about flows and network as it is about boundaries, and this point is important for what follows. Where consideration of the intrinsic value of marine environments is concerned, an appreciation of socio-spatial complexity is especially important: marine environments account for the deepest waters over which no sovereign nation has legal, political or economic control but in which many have extensive interests – they are part of the global commons – owned by none, to be nourished by all. They include territorial waters from the 200 nautical mile mark to the outer edge of coastal zones, and function to reinscribe national, regional and local allegiances. They embrace these coastal zones, marked by above-average concentrations of people and economic activity (NetCoast 2001). They neither respect nor respond to the boundaries imposed on them by cartography. What flows from river systems to estuaries, from territorial waters to deep seas to circulate around the globe in water and air, recognises neither juridical nor jurisdictional boundaries, structures and processes.

This lack of recognition raises questions of governance to which social scientists must pay heed in addressing the implementation deficit. Wapner (1995: 45) suggests that global efforts on behalf of the environment demand world order reform that 'enlarges the political imagination and expands the conceptual boundaries of future thinking and possible action with regard to environmental issues'. In this light, different camps of supra-statists variously argue the need to foster *world government*. However conceived,

> world government may simply be the worldwide legitimation and further codification of the state-system. Critics also point to its infeasibility . . .
> There is no reason to believe that a world government would necessarily be more benign than existing state governments [and] . . . there is nothing intrinsic to world government that precludes further ecological decay. (Wapner 1995: 57)

Alternatively, advocates of sub-statism argue for collapsing nationstates and decentralising political authority to systems of *local governance* on the grounds that centralisation demands and underwrites the technologies of super-industrialisation and economic globalisation. Attending this transformation of production and consumption are massive disruptions to ecological processes and alternative practices of social and economic exchange. Centralisation also overwhelms the capacity to be engaged and empowered.

Sub-statism is a response to these dilemmas, and among its variants is a common emphasis on willingly *living-in-place* in order to be ecologically sensitive, and a claim that coercive global government will not achieve these ends. Wapner (1995) nevertheless concludes that a focus on the parts does not guarantee the stewardship of the whole.

In the final analysis, both supra-statists and sub-statists acknowledge that states have addressed environmental degradation via national legislative frameworks, other command-and-control mechanisms, and intergovernmental cooperation with the United Nations, the World Bank, the Global Environment Facility, and so forth. Limits to success suggest that 'the forces that cause environmental degradation continue unabated and have in fact gained momentum over the past twenty-five years – the period marking the heyday of international environmental efforts' (Wapner 1995: 47). In short, the implementation deficit continues.

KEY DEBATES

At least three debates circulate around the question of whether the implementation deficit is attributable to tensions between the global and the local. The first centres on metaphors to describe sustainable development and uphold its position inside the logic of capitalism. The second differentiates between globalising from above (perpetuating the status quo) and the counter-movement of globalising from below (questioning that same system). The third concerns the significance of global and local scales for the commons – and marine environments in particular.

Sustaining capitalism?

At its most basic, sustainable development is managing a *triple bottom line* between economy, society and environment using two strategies. One relies on laws and regulations (related to shipping, coastal development or deep-sea trawling, for example). The other depends on participatory devices (such as government–community partnerships for coastcare, wetlands preservation or coral reef protection). Hart (1999) unsettles the apparent *evenhandedness* of this model because the economic is privileged *in fact*. For her, 'the economy exists entirely within society . . . [which] exists entirely within the environment [which] surrounds society [and] because people need food, water and air to survive, society can never be larger than the environment'. Her nested understanding of sustainable development reflects ideas of relational scale (Howitt 2002), which underscores the reliance of the economic on the social, of the social on the environmental. It also invokes ideas of scalar relations (Howitt 2002), implying that people must remember their place in the environment – in that which surrounds and nourishes them.

Sustainable development may also be understood as a strategy to safeguard the *human ecosystem* (Machlis et al. 1997) in which critical natural, socio-economic and cultural resources interact with general social structures that regulate their use. Social structures encompass social *institutions* (health, justice, commerce, education, government and so on); social *cycles* (physiological, individual, institutional and environmental); and social *order* (identity, norms and hierarchies). The system is an arrangement of 'biophysical and social factors capable of adaptation and sustainability over time [that can] . . . be described at several spatial scales' (Machlis et al. 1997: 351). This reference to scale is important because no part of the human ecosystem exists outside spatial frameworks or the flows and boundaries that delimit them.

Sustainable development is also viewed as a strategy to value and accumulate natural, human, social, physical, fiscal and financial, and organisational assets (Pretty & Frank 2000; see also Stratford & Davidson 2002). Where an appreciation exists of the integratedness of the *capital assets* that comprise social-life-in-place, and where there are well-developed capacities for civic participation and political engagement, the preconditions to overcome the implementation deficit seem strong, and the tendency to deplete the stock of assets less likely.

Despite the intellectual usefulness of these metaphors, they perpetuate sustainable development's position 'inside' the logic of capitalism. People thus find it difficult to implement *at any scale along the continuum from local to global* the key principles of sustainability necessary to transform institutions and organisations, and to foster nourishing capacities of being-in-place. Indeed, much of the early radical potential of environmentalism (Schlosberg & Dryzek 2002) remains unfulfilled; sustainable development is 'business as usual' (Davidson 2000). This outcome privileges the logic of economic growth inside the framework of ecological modernisation.

Ecological modernisation

Earlier I referred to the debate about globalisation from above and below. Ecological modernisation assumes that super-industrialisation will produce the means by which to protect the environment as it produces the goods and services to advance quality of life. In practice, this idea of quality of life is highly modernised and westernised; it also upholds the status quo.

The values of ecological modernisation influence global thinking and local action for sustainable development from above. At the supra-national scale, for instance, various environmental conferences and commissions have been established by the United Nations. The institutional history and critiques of this system are well documented (WCED 1987; United Nations 1992a, 2002; Doyle 1998; UNESCO 2003).

One of the foundational figures of environmental sociology, Buttel (2003: 329) argues that these efforts have produced a 'hopeful pattern of international collaboration and agreement that has subsequently become one of the pillars of modern thought about how a more promising environmental future can be made possible'. He views the logic behind an international approach to environmental reform as compelling. It has multiplying effects at various relational and jurisdictional scales involving many stakeholders. It is an alternative to command-and-control mechanisms, a route to policy-making that may be more egalitarian and inclusive.

Buttel (2003) is suspicious of ecological modernisation and the faith that certain adherents have in market, state and private action. Rather, he underscores how networks, alliances and coalitions in various locations and at many scales have championed 'globalisation from below' via environmental justice, social justice and the civil rights movements that blend 'the themes of environmentalism and social and racial justice in a way that can bring forward an impressive level of mobilization around local and regional environmental issues' (Buttel 2003: 313). Although he does not specifically refer to alliances whose focus is marine environments, there are many, and most tie into other networks that operate simultaneously at *multiple scales* (Ecological Internet Incorporated 2004).

What's local, what's global?

The idea that different scales work concurrently informs the third debate about the implementation deficit (and governance of marine environments more specifically). Many examples show that coastcare, rivercare, marine and land-based projects bring together community members and groups, private enterprise and government in the local management of marine environments. Efforts are often supported by grants, subsidies and philanthropic endowments, and the gains that are sought are often ecologically modernising. For example, many projects require a focus on the next grant, the next endowment; on ensuring the flow of capital necessary to maintain the project and – somewhere upscale – to satisfy the interests of large companies, property-owners and shareholders seeking to maximise their investments from afar. Sometimes the result is suboptimal: in Australia, for example, the part-sale of Telstra, the Commonwealth-owned telecommunications carrier, to fund the Natural Heritage Trust for coastcare, landcare and related projects did not provide anticipated returns to investors.

There are many instances, too, of critical and radical actions designed to question and sidestep, unsettle and eventually erode the practices of economic globalisation (Starr & Adams 2003). Among the actions for the aquatic environment are those staged by organisations such as Greenpeace to conserve species and habitats, or by local community groups in underdeveloped regions to protect access to marine resources for subsistence livelihoods (Ecological Internet Incorporated 2004; see also Nichols 1999; Cole 2003).

The degradation of marine environments continues apace, notwithstanding the advent of local actions engaged with the state and capital (and thus inside sustainable development and the logic of capital) and those attempting to stand outside or supplant these institutions with forms of sustainable development beyond capitalism.

Perhaps the ongoing and increasing threat to marine environments indicates that the implementation deficit has less to do with whether local action is 'inside' or 'outside' the system and more to do with the dilemmas of scale per se. Some commentators, such as McLaren (2001), are voluble in their claims about the importance of local engagement and empowerment as an antidote to the effects of economic globalisation. Others suggest that it is important not to over-invest the local with the status of 'miracle scale' in relation to sustainable development. Gibbs & Jonas (2000), for example, are impatient with aspects of the argument that the local is the most appropriate site for environmental policy interventions just because it is circuitously described as the level of action that is central to sustainable development by national and, more particularly, by international conventions. For them, as for me, the rhetorical construction of the local is inherently problematic because it does not acknowledge the complexities of flow, seeking to 'fix' spatial categories where none, in practice, exist.

Assuredly, global thought and actions for marine environments are perhaps more apparent than real and appear to have had little overall effect on, for example, species or habitat decline, the loss of coastal lands to urban development, or international piracy and trafficking. Indeed there is a veritable industry to address the global dimensions of managing human impacts on marine environments that is based around international preparatory committees, conferences, conventions and agreements. Some might suggest these organisations add little to poverty alleviation and environmental protection, and much to the profit margins or performance claims of airlines, hotel chains, universities and research and development organisations, governments, and so on. Others would counter that to do nothing is unconscionable.

The United Nations exemplifies how diverse local–global tensions typify the governance of marine environments and partly explain the implementation deficit. Chapter 17 of *Agenda 21* is entitled 'Protection of the oceans, all kinds of seas, including enclosed and semi-enclosed seas, and coastal areas and the protection, rational use and development of their living resources'. Seven program areas are advanced: (a) integrated management and sustainable development of coastal and marine areas, including exclusive economic zones; (b) marine environmental protection; (c) sustainable use and conservation of marine living resources of the high seas; (d) sustainable use and conservation of marine living resources under national jurisdiction; (e) addressing critical uncertainties for the management of marine environments and climate change; (f) strengthening international, including regional, cooperation and coordination; and (g) sustainable development of small islands (United Nations 1992b: para. 17.1).

Each program area demands attention to scale. To return to earlier discussions about how marine environments neither respond to nor respect juridical and judicial boundaries, it is important to add that existing structures and processes of governance cannot yet address the complexities that this environment presents. Taking program area (a) as an example, *integrated* management of coastal and marine environments up to the edge of the borders of territorial boundaries (exclusive economic zones) is unlikely without due regard for both land-based activities and those in the deep seas. Nevertheless, the rhetoric of global governance constitutes the solution touted in program area (a) as spatially contained between the shoreline and the 200 nautical mile mark. Certainly, local activists and global policy-makers are addressing the limitations of this rhetoric (see National Oceanographic and Atmospheric Administration 2004). Nonetheless, the nature of funding is political and sectoral (which does not bode well for the radical integration required to manage human activities in relation to marine environments). Furthermore, many projects are not supported by recurrent budgets and cannot compete with economic activities that are.

Integration is especially important for marine environments, and the failure to recognise the arbitrariness of the local and global is captured by the observation that:

> In dealing with the ocean we are forced to face nearly all [integrative] problems of war and peace, security and economy . . . Arvid Pardo in 1967 also specified this through his basic seminal idea that 'all aspects of ocean space are inter-related and should be treated as a whole'. The role of the ocean and coasts in the new global service-oriented economy and the globalization process amply demonstrates the need for an adequate ocean governance. The legal, international agreements are in place to achieve this. (Kullenberg 2002: 774)

Kullenberg's pronouncements were supported at the UN's World Summit on Sustainable Development (WSSD) in September 2002 in Johannesburg. Paragraphs 30–36 of the *Plan of Implementation* (United Nations 2002) refer to oceans, coasts and islands, and are drawn from a larger section (IV). That section is entitled 'Protecting and managing the natural resource base of economic and social development' and it privileges the environment's instrumental values for humanity rather than its intrinsic worth.

Table 9.1 summarises targets and timetables for marine environments adopted at the WSSD. The language used illustrates how the sustainable development framework continues to promote ecologically modernising understandings of environmentalism per se, remains enmeshed in the logic of capitalism, and is characterised by various scalar ambiguities and slippages related to locale, region, nation, and international domains.

'Show me the money'

In October 2003, the WSSD targets and timetables for marine environments were the centrepiece of the second Global Forum on Oceans, Coasts, and Islands in Paris; the first had been held in 2001 and had ensured that marine environments and coastal and island peoples were not neglected in the WSSD process in Johannesburg.

I was an independent observer at the October forum. In two days of intensive pre-conference meetings and three days of formal proceedings, delegates were to (a) focus on how to implement the WSSD targets and timetables outlined in Table 9.1; (b) report on what they and their sectors had been and would be doing to advance the protection of the oceans, coasts and islands; (c) discuss gaps in partnership initiatives from WSSD and forge new means to close those; (d) examine the

Table 9.1 *Major Targets and Timetables Adopted at the World Summit on Sustainable Development on Oceans, Coasts and Islands*

Integrated ocean and coastal management
- Encourage the application of the ecosystem approach by 2010 for the sustainable development of the oceans, particularly in the management of fisheries and the conservation of biodiversity.
- Establish an effective, transparent and regular inter-agency coordination mechanism on ocean and coastal issues within the United Nations system.
- Promote integrated coastal and ocean management at the national level and encourage and assist countries in developing ocean policies and mechanisms on integrated coastal management.
- Assist developing countries in coordinating policies and programs at the regional and subregional levels aimed at conservation and sustainable management of fishery resources and implement integrated coastal area management plans, including through the development of infrastructure.

Fisheries
- Implement the FAO International Plan of Action to Prevent, Deter and Eliminate Illegal, Unreported and Unregulated Fishing by 2004.
- Implement the FAO International Plan of Action for the Management of Fishing Capacity by 2005.
- Maintain or restore depleted fish stocks to levels that can produce their maximum sustainable yield on an urgent basis and where possible no later than 2015.
- Eliminate subsidies that contribute to illegal, unreported and unregulated fishing and to overcapacity.

Conservation of biodiversity
- Develop and facilitate the use of diverse approaches and tools, including the ecosystem approach, the elimination of destructive fishing practices, the establishment of marine protected areas consistent with international law and based on scientific information, including representative networks by 2012.

Protection from marine pollution
- Advance implementation of the Global Programme of Action for the Protection of the Marine Environment from Land-based Activities in the period 2002–2006 with a view to achieve substantial progress by 2006.

Science and observation
- Establish a regular process under the United Nations for global reporting and assessment of the state of the marine environment, including socioeconomic aspects, by 2004.

Small Island Developing States
- Develop community-based initiatives on sustainable tourism in small island developing States by 2004.
- Reduce, prevent, and control waste and pollution and their health-related impacts in small island developing States by 2004 through the implementation of

Table 9.1 (*cont.*)

the Global Programme of Action for the Protection of the Marine Environment from Land-based Activities.
• Support the availability of adequate, affordable and environmentally sound energy services for the sustainable development of small island developing States, including through strengthening efforts on energy supply and services by 2004.
• Undertake a comprehensive review of the implementation of the Barbados Programme of Action for the Sustainable Development of Small Island Developing States in 2004.

(*Source:* Global Forum on Oceans, 2003: 1)

particular needs of small island developing states (SIDS) in the lead-up to another international meeting in Mauritius in August 2004 at which the 1994 Barbados Plan of Action on SIDS will be revisited; (e) identify and discuss new issues challenging sustainable development for marine environments; and (f) pinpoint ways to better involve the private sector in the implementation of the WSSD (Global Forum on Oceans, Coasts and Islands 2003).

This agenda was ambitious and thoroughly enmeshed in networks of influence, with prominent delegates from the Global Environment Facility, the World Bank, and the United Nations and its affiliate organisations under pressure to promise *recurrent* resources to fund local actions that advance global thinking on sustainable development and marine protection. It was perhaps ironic, then, that on the first of the pre-conference days, delegates visited and revisited the WSSD timetable for oceans, coasts and islands, often making minimal reference to how that agenda was to be financed. This apparent lack of attention to the 'real' bottom line led one experienced player to utter what became a waggish conference catch-cry: 'show me the money.'

These words are a timely reminder that sustainable development is inside local, sub-national, national, regional and international monetary systems. It is inside bureaucracies and political systems typified by competing demands, interests and ideological priorities. It is inside political and fiscal calculations about what is politically expedient and practical within existing regimes to implement sustainable development's principles. Equally, it is clear that experienced personnel from international donor groups are well aware of the financial and capacity constraints on implementing the rhetoric of sustainable development, despite the fact that during 'its first decade, the GEF allocated $4.2 billion in grant financing, supplemented by more than $12 billion in additional financing, for 1,000 projects in 160 developing countries and countries in economic transition' (Global Environment Facility 2001).

FUTURE DIRECTIONS

My work has been informed by a commitment to *being in place*, a capacity that I think will foster democratic and civic engagement and empowerment to nourish that which is simultaneously 'ours' and 'not ours'. Sitting among hundreds of delegates to the Second Global Forum in Paris in October 2003, I was struck by a number of sensations: the international community's passion for the governance of marine environments; the sense of urgency about the task to fulfil the impossibly difficult WSSD timetable of action; deep differences of opinion about the scale or scales at which to effect change; common commitment for oceans, coasts and islands. Nevertheless, the overwhelming tenor resembled what Schnaiberg (1980) called the 'treadmill of production', which may serve to underscore my contention that while sustainable development remains embedded in capitalism and ecological modernisation strategies both the general implementation deficit and the untenable exploitation of marine environments will persist and accelerate.

Discussion Questions

1. In what ways are ideas of scale used in environmental sociology?
2. Wapner (1995) suggests that supra-statism and sub-statism remain inside 'statist' thinking, although they function as critiques of it. Do you agree? Why or why not?
3. The focus of this chapter has been on oceans as a part of the world's 'commons'. Is this type of analysis applicable to private property and sustainable development as well?
4. How would you define 'global' and 'local'?
5. What do you understand to be the major differences between globalisation from above and below?
6. The principle of integration is central to sustainability. What does the term mean?
7. Is the implementation deficit misnamed? Isn't the 'fuss' merely a debate about incremental – as distinct from radical – change?
8. Is economic globalisation different from globalisation generally? How? What significance might this difference have for thinking about sustainable development?
9. Speculate: would long-term recurrent funding, such as a sustainability tax levy, help overcome the limited successes of the past?
10. Marine environments do not respect or respond to the sorts of boundaries that human beings impose on them. Discuss.

Glossary of Terms

Anthropocentric: human-centred; in environmentalism, assuming the central importance of people's welfare in relation to the welfare of other species or habitats and ecosystem processes.

Anthropogenic: human-induced and/or human-sourced change.

Artisanal: as distinct from commercial; small-scale ventures typically involving low-level, craftly approaches rather than sophisticated and industrial ones.

Benthic: of the ocean floor; from the Greek 'depth of the sea'; the term describes flora and fauna that are fixed to or that dwell on the sea bottom (Macquarie Dictionary 1985).

Carrageenan: one of the family of hydrocolloids that includes gelatin, pectin, xantham gum, agar, gellan, locust bean gum and Carboxyl Methyl Cellulose. Carrageenan is derived from a number of tropical and coldwater species of seaweed, and is refined to a white powder for addition to a wide variety of food, cosmetics and pharmaceuticals as an emulsifier, stabiliser and thickener (Bixler 1996).

Globalisation   is seen as the 'intensification, widening and deepening, of international networks across the economic, military, technological, ecological, migratory, political and cultural flows. That definition . . . argues that the intensification of international networks, leading to '*interconnectedness*', is unique to the contemporary period, but stresses that the networks are created as part of an ongoing process' (Cole 2003: 79; original emphasis).

Hegemony   involves the naturalisation of particular forms of power through intellectual and moral leadership, the production of 'authority' and the use of the 'majority' and 'consensus' to uphold the privilege of those who exercise those forms of power.

Instrumentalist: in relation to philosophy, this term describes the condition of instrumentalism 'which maintains that . . . ideas have value according to their function in human experience or progress' (Macquarie Dictionary 1985).

Mariculture: cultivation of foods from marine sources and environments.

Nautical mile: equal to 1852 metres, this unit of measurement is used in both marine and aeronautical navigation, and was originally defined as one minute of latitude (Macquarie Dictionary 1985).

Subsidiarity: the process of devolving to the most appropriate level of governance a particular responsibility or responsibilities.

References

Armstrong, D., and E. Stratford (in review), 'Partnerships for local sustainability and local governance in a Tasmanian settlement', *Local Environment*.
Bixler, H.J. 1996, 'Recent developments in manufacturing and marketing carrageenan', *Hydrobiologia* 326–327: 35–57.

Burges Watson, D., and E. Stratford (in preparation), 'Engendering microbi-cides: the risky geopolitics of CarraguardTM and HIV/AIDS prevention.'

Buttel, F.H. 2003, 'Environmental sociology and the explanation of environ-mental reform', *Organization and Environment* 16(3): 306–44.

Cole, H. 2003, 'Contemporary challenges: globalisation, global interconnect-edness and that "there are *not* plenty more fish in the sea". Fisheries, governance and globalisation: is there a relationship?' *Ocean and Coastal Management* 46: 77–102.

Costanza, R. 1999, 'The ecological, economic, and social importance of the oceans', *Ecological Economics* 39(2): 199–213.

Crowley, K. 1999, 'Explaining environmental policy: challenges, constraints and capacity'. In K.J. Walker and K. Crowley, *Australian Environmen-tal Policy 2: Studies in decline and devolution*, Sydney: UNSW Press, pp. 45–64.

Davidson, J. 2000, 'Sustainable development: business as usual or a new way of living?' *Environmental Ethics* 22(1): 25–42.

Doyle, T. 1998, 'Sustainable development and Agenda 21: the secular bible of global free markets and pluralist democracy', *Third World Quarterly* 19(4): 771–86.

Eagle, J., and H.T.J. Barton 2003, 'Answering Lord Perry's question: dissecting regulatory overfishing', *Ocean and Coastal Management* 46: 649–79.

Ecological Internet Incorporated 2004, *Eco-Portal: The Environmental Sus-tainability Info Source, Ecological Internet Incorporated.* http://www.eco-portal.com/Ocean/Organizations/Campaigns/welcome.asp.

Fallon, L.D., and E. Stratford 2003, *Issues of Sustainability in the Southern Oceans Fisheries: The case of the Patagonian toothfish*, Hamburg: Lighthouse Foun-dation and the University of Tasmania.

Foster, E.G., and M. Haward 2003, 'Integrated management councils: a con-ceptual model for ocean policy conflict management in Australia', *Ocean and Coastal Management* 46: 547–63.

Friedheim, R.L. 1999, 'A proper order for the oceans: an agenda for the new century'. In D. Vidas and W. Ostreng, *Order for the Oceans at the Turn of the Century*, The Hague/London/Boston: Kluwer Law International, pp. 537–57.

Future Harvest 2003, 'Global demand for fish rising: fish farming is the fastest growing field of agriculture', *Future Harvest.* http://www.futureharvest.org/earth/fish.html.

Gamble, D.N., and M.O. Weil 1997, 'Sustainable development: the challenge for community development', *Community Development Journal* 32(3): 210–22.

Ghina, F. 2003, 'Sustainable development in small island developing states', *Environment, Development and Sustainability* 5(1–2): 139–65.

Gibbs, D., and A.E.G. Jonas 2000, 'Governance and regulation in local envi-ronmental policy: the utility of a regime approach', *Geoforum* 31(3): 299–313.

Global Campaign for Microbicides 2003, Global Campaign for Microbicides. http://www.global-campaign.org/.

Global Environment Facility 2001, *GEF Global Action on Water: A decade of managing transboundary waters*, Washington DC: GEF.

Global Forum on Oceans, Coasts and Islands 2003, Pre-Conference and Conference Programme. http://www.globaloceans.org/globalconference/about.html#purposes.

Halfara, J., and R.M. Fujitab 2002, 'Precautionary management of deep-sea mining', *Marine Policy* 26: 103–6.

Hart, M. 1999, *Guide to Sustainable Community Indicators*, Andover MA: Hart Environmental Data.

Howitt, R. 2002, 'Scale and the other: Levinas and geography', *Geoforum* 33: 299–303.

Huber, M.E., R.A. Duce, J.M. Bewers, D. Insull, J. Ljubomir and S. Keckes 2003, 'Priority problems facing the global marine and coastal environment and recommended approaches to their solution', *Ocean and Coastal Management* 46: 479–85.

Jones, A.T., and C.L. Morgan 2003, 'Code of practice for ocean mining: an international effort to develop a code for environmental management of marine mining', *Marine Georesources and Geotechnology* 21(2): 105–14.

Joost, D.G., A. Burton-James and G. Cambers 2002, 'Wise Practices for Conflict Prevention and Resolution in Small Islands'. Results of a Workshop on Furthering Coast Stewardship in Small Islands, Dominica, 4–6 July 2001. Coastal Region and Small Island Paper 11. Paris: UNESCO.

Kullenberg, G. 2002, 'Regional co-development and security: a comprehensive approach', *Ocean and Coastal Management* 45: 761–76.

Kupke, V. 1996, 'Local Agenda 21: local councils managing for the future', *Urban Policy and Research* 14(3): 183–98.

Lindeboom, H. 2002, 'The coastal zone: an ecosystem under pressure'. In J.G. Field, G. Hempel and C.P. Summerhayes, *Oceans 2020: Science, trends and the challenge of sustainability*, Washington DC: Island Press, pp. 49–84.

Machlis, G.E., J.E. Force and W.R. Burch 1997, 'The human ecosystem Part I: the human ecosystem as an organizing concept in ecosystem management', *Society and Natural Resources* 10: 347–67.

McConnell, M. 2002, 'Capacity building for a sustainable shipping industry: a key ingredient in improving coastal and ocean and management', *Ocean and Coastal Management* 45(9): 617–32.

McLaren, D. 2001, Guest Editorial. 'From Seattle to Johannesburg: "anti-globalisation" or "inter-localism"?' *Local Environment* 6(4): 389–91.

Mitchell, C.L., and L.O. Hinds 1999, 'Small island developing states and sustainable development of ocean resources', *Natural Resources Forum* 23(3): 235–44.

National Oceanographic and Atmospheric Administration 2004, 'White Water to Blue Water Initiative: A partnership to Link Freshwater and Oceans'. http://www.international.noaa.gov/ww2bw/.

NetCoast 2001, 'A Guide to Integrated Coastal Zone Management, NetCoast Netherlands'. http://www.netcoast.nl/info/coast.htm.

Nichols, K. 1999, 'Coming to terms with integrated coastal management: problems of meaning and method in a new arena of resource regulation', *Professional Geographer* 51(3): 388–99.

Pallis, M. 2002, 'Obligations of states towards asylum seekers at sea: interactions and conflicts between legal regimes', *International Journal of Refugee Law* 14(2–3): 329–64.

Pelling, M., and J.I. Uitto 2001, 'Small island developing states: natural disaster vulnerability and global change', *Global Environmental Change Part B: Environmental Hazards* 3(2): 49–62.

Pirtle, C.E. 2000, 'Military Uses of Ocean Space and the Law of the Sea in the New Millennium', *Ocean Development and International Law* 31(1): 7–45.

Population Council 2003, 'HIV/AIDS Microbicides'. http://www.popcouncil.org/hivaids/microbicides.html.

Pretty, J., and B.R. Frank 2000, *Participation and Social Capital Formation in Natural Resource Management: Achievements and lessons*, International Landcare Conference: 'Changing Landscapes – Shaping Futures', Melbourne.

Schlosberg, D., and J.S. Dryzek 2002, 'Political strategies of American environmentalism: inclusion and beyond', *Society and Natural Resources* 15: 787–804.

Schnaiberg, A. 1980, *The Environment: From surplus to scarcity*, New York: Oxford University Press.

Smith, H.D. 2003, Editorial, 'Emerging issues in oceans, coasts and islands', *Marine Policy* 27: 289–90.

Starr, A., and J. Adams 2003, 'Anti-globalization: the global fight for local autonomy', *New Political Science* 25(1): 19–42.

Stratford, E., and J. Davidson 2002, 'Capital assets and intercultural borderlands: socio-cultural challenges for natural resource management', *Journal of Environmental Management* 66: 429–40.

Summerhayes, C.P., J.G. Field and G. Hempel 2002, Introduction. In J.G. Field, G. Hempel and C.P. Summerhayes (eds), *Oceans 2020: Science, trends and the challenge of sustainability*, Washington DC: Island Press, pp. 1–8.

Trist, C. 1999, 'Recreating ocean space: recreational consumption and representation of the Caribbean marine environment', *The Professional Geographer* 51(3): 376–87.

UNESCO 2002, *The MAB Biosphere Reserves Directory, Australia – Macquarie Island*. http://www2.unesco.org/mab/br/brdir/directory/biores.asp?mode=all&Code=AUL+03.

UNESCO 2003, *Man and the Biosphere Program*, UNESCO. www.unesco.org/mab/.

United Nations 1992a, *Agenda 21*, UN Conference on Environment and Development.

United Nations 1992b, *Agenda 21*, Chapter 17, 'Protection of the oceans, all kinds of seas, including enclosed and semi-enclosed seas, and coastal areas and the protection, rational use and development of their living resources', Rio de Janeiro: UN.

United Nations 2002, *World Summit on Sustainable Development Johannesburg Plan of Implementation*. http://www.johannesburgsummit.org/.

Vanderzwaag, D. 2002, 'The precautionary principle and marine environmental protection: slippery shores, rough seas, and rising normative tides', *Ocean Development and International Law* 33(2): 188–212.

Wapner, P. 1995, 'The state and environmental challenges: a critical exploration of alternatives to the state-system', *Environmental Politics* 4(1): 44–69.

Wells, P.G., R.A. Duce and M.E. Huber 2002, 'Recent development. Caring for the sea – accomplishments, activities and future of the United Nations GESAMP (the Joint Group of Experts on the Scientific Aspects of Marine Environmental Protection)', *Ocean and Coastal Management* 45: 77–89.
Women's Global Health Imperative 2003, 'AIDS has a woman's face: gender and power: new strategies for HIV prevention'. http://hivinsite.ucsf.edu.
WCED (World Commission on Environment and Development) 1987, *Our Common Future*, Melbourne: Oxford University Press.

CHAPTER TEN

CITIZENSHIP AND SUSTAINABILITY
Rights and Responsibilities in the Global Age

Julie Davidson

As an object of academic and political concern, the concept of citizenship has received ever-increasing attention with enlargement of the political context from the nation-state to the globe over the last several decades. The Australian government, for example, commissioned a report into citizenship and civic education out of concern for rising political apathy (Civic Expert Group 1994). The volume of scholarly publications on the subject has accelerated since the 1980s. Such theorising occurs against a background of substantive change in political communities and political legitimacy effected by an upswing in globalising processes; structural crises in the world economic system; increasing global risks and uncertainties; growing inequalities among rich and poor both internationally and intranationally; environmental degradation and decline at both local and global scales; and greater ethnic diversity and multiculturalism in formerly homogeneous populations. All of these shifts portend increasing global interdependence and have considerable implications for how the various dimensions of citizenship are perceived but most particularly for the apparent rights and responsibilities adhering to citizenship.

In the context of such shifts, two discourses are uppermost in shaping the nature of citizenship. These are the discourses of globalisation and sustainability, both of which are contested. On the one hand, globalisation is characterised as 'the beginning of a new epoch in human affairs' (Held et al. 1999: 494) or as a transition from one state to another (Albrow 1996). For Immanuel Wallerstein (2000), it is an era of systemic crisis in the system of wealth production and the political legitimacy that supports it, while for Martin Albrow (1996: 85), globalisation is best employed as 'a marker for a profound social and cultural transition', and most fruitfully conceived as the global age, an era when the global rather than the modern provides its defining features. Beck (1992), on the other hand, interprets globalisation as another phase of modernisation beyond simple modernity where risks become globalised and the principles underpinning wealth production come under scrutiny. However globalisation is conceived, the shift to the global as primary referent is of such a magnitude that many of the organising concepts of modernity are losing their purchase on the democratic imagination with significant implications for the traditional concept of citizenship.

Parallel with the discourse around globality and globalisation, and seemingly at odds with it, is the discourse of sustainability, with economic forces driving the former. Current patterns of economic growth have been described as unsustainable on the basis that they are 'environmentally disruptive, macroeconomically counterproductive, and socially divisive' (Spangenberg 2001: 32). The transformations in systems of economic and political governance, knowledge and values that will be needed to enable the principles of sustainability to be enacted across the globe are perceived to be so far-reaching that the basic organising concepts of modernity have to be rethought, including the nation-state, liberal democracy, and citizenship. However, while there is broad agreement that past industrial practices have disrupted ecosystems and global processes such as climate regulation, how sustainability should be pursued is strongly disputed. Some interests maintain that sustainable development is best implemented through technological and managerial innovation, with environmental protection traded off against economic growth. At the other end of the sustainability continuum are those who interpret sustainability as an ethic that forms the basis for restructuring human productive activity and its relationship to nature (Jacobs 1995; Davidson 2000).

Retheorising citizenship in the context of globalisation began in earnest around 1990, with the relationship between citizenship and ecology first being recognised in Bart van Steenbergen's (1994b) concept of the *ecological citizen*. Even at this nascent stage of its theorisation, the global context of ecological citizenship was acknowledged and outlined. In this chapter ecological citizenship is discussed in the context of global agendas of social, economic and environmental improvement but also from the perspective of retheorising citizenship more generally. Thus the relationships between concepts of obligation and responsibility, active citizenship and its ecological and obligate versions, stewardship, and global or cosmopolitan citizenship are explored with reference to sustainability.

The primary aim of this chapter is to explore recent debates around citizenship as they relate to the global environmental crisis in order to advance an understanding of citizenship for sustainability. In what follows, the global transition forms the context for recent citizenship debates, while globalisation is recognised as a key driver behind observed changes to this central institution of modern life. Next, the implications of economic globalisation for citizenship and its supporting institutions are outlined, from whence discussion proceeds to a synopsis of how citizens' rights are changing, of the shift to a cosmopolitan paradigm, and thence to a discussion of the relationship between citizenship and sustainability. Of particular note here are concepts such as planetary citizenship, and notions of obligation and responsibility as they are propounded in neo-liberal and ecological conceptions of active citizenship. Lastly, future prospects and future research are canvassed.

BACKGROUND TO THE ISSUES

Citizenship has been one of the central organising concepts of modern democratic political systems. Its main components, developed over the several centuries since the liberal revolutions of the 17th and 18th centuries, were captured and theorised by T. H. Marshall (1950), whose work has been the standard reference on liberal-democratic citizenship for the last half-century. Marshall identified three sets of rights whose evolution began with the recognition of *civil* rights: freedom of speech, thought and religion; the right to own property; and the right to justice. These were then supplemented by *political* rights of participation in decision-making processes through voting, and finally by a *social* component that ensured rights to economic welfare sufficient for full participation in social life. While there has been much criticism of Marshall's model of citizenship, Turner (2001: 190) believes that his main contribution to citizenship theory is the claim of a 'permanent tension between the principles of equality that underpin democracy and the de facto inequalities of wealth and income that characterise the capitalist market-place', the function of citizenship being to mitigate the negative effects of inequity by redistributing wealth on the basis of rights. As we shall see, citizenship is losing this capacity and in so doing the inadequacies of the Marshallian model are becoming plainly evident, as are the inherent contradictions of the market economy.

In the context of changes to citizenship, the end results have been encapsulated in diverse ways as its decline or marginalisation (Falk 2000), its erosion (Turner 2001), and its decomposition (Cohen 1999), although the urgency of promoting good citizenship is disputed by some

(Kymlicka & Norman 1995). As well, Cohen (1999) argues that it is as a consequence of broad political and economic trends that the modern model of citizenship has been rendered 'anachronistic' and that there has arisen a renewed interest in the relationship between democracy, justice and identity, which until recently had very neatly mapped onto each other in the setting of the nation-state.

Before discussing the changes that citizenship is undergoing, their political, social, economic, and ecological contexts will be elaborated since such changes cannot be understood apart from the transformations occurring in these contexts, transformations that signify far-reaching consequences for citizenship conceived as a status position.

Why citizenship is changing

Among theorists, there is considerable consensus on the effects of economic globalisation on nationally based political and economic systems, exemplified by Falk (2000: 6) thus:

> The essential argument is that economic globalization is weakening territorial ties between people and the state in a variety of ways that are shifting the locus of political identities, especially of elites, in such a manner as to diminish the relevance of international frontiers, thereby eroding, if not altogether undermining, the foundations of traditional citizenship.

The primary transformation paralleling the weakening of ties is the delinking of nation from state (Albrow 1996), driven by the weakening of political power vis-à-vis economic power (Altvater 1999).

However, even as traditional citizenship is apparently being undermined, growing interconnection among the world's peoples is fostered by the acceleration of global flows of people, goods, ideas, and information on the one hand, and the proliferation of trans-boundary problems on the other (Held 1999, 2000). Such an upsurge in 'environmentally dangerous circumstances' (Falk 2001: 222) is perceived as being responsible for mounting insecurity and environmental inequity (Elliot 2002).

One way in which the meaning of citizenship is changing as the nation-state is being supplanted in some of its roles is in the protection of rights. Previously, the rights of citizens tied to individual nations were sufficient to protect human rights. However, as connections across the globe have multiplied, it has become clear that the 'concept of human rights is more global than that of citizenship in so far as it encompasses notions of entitlement that transcend considerations of nationality' (Dean 2001: 493). Human rights have gone global and consequently the authority of national institutions in protecting basic rights is being superseded by international institutions, such as international agreements and supra-national courts (Cohen 1999; Held 1999).

KEY DEBATES

There is no doubt that recent global transformations have impacted on citizenship and its supporting institutions: courts, parliament, councils of local government, education and health systems, and social services. Generic change in the substance of citizenship has been the topic of a number of scholarly works. Thus for Bart van Steenbergen (1994b), the meaning of citizenship is in flux, as is the meaning of participation, while for others, such as Dean (2001) and Cohen (1999), its practice is assuming greater importance than its status. Similarly, Herman van Gunsteren (1994) encapsulates the shift from a rights focus to a responsibility focus as a move from the 'practices of liberation' (entitlement) to the 'practices of freedom' (competence, and the sensible and responsible use of freedom). Citizenship is thus thought to be entering a new paradigm accompanied by new forms of political organisation (Cohen 1999).

Among the key issues occupying theorists and governments in the new paradigm are questions of the right to a healthy environment, the rights of nature, the efficacy of a global civil society, the ground of citizenship, the responsibilities of citizenship, and the quality of citizens. The following sections canvass these issues.

Are rights still relevant?

Changes in the sphere of citizenship have been particularly evident in the context of rights. Citizenship has always been understood as a constellation of rights and responsibilities, although in modern times theory and practice have concentrated on the rights component. More recently, the ecological crisis has generated a greater concern with responsibilities of citizenship. Nonetheless, there have been significant changes in rights elements both nationally and globally. Within the *civil* component, the right to justice has become as much a human right as a civil right, while the absolute rights of property ownership are being continuously modified by legislatures, by regulatory agencies, by zoning practices, and by the decisions of planning tribunals (Sagoff 1988). Thus, as human rights have become an international concern, so too are transnational companies, which also possess the attributes of private individuals, being challenged to act with greater regard for the consequences of their decisions through codes of corporate responsibility (Soares 2003).

Paradoxically, as responsibility has emerged as an increasingly more important component of citizenship, the *political* rights that would allow citizens to exercise more responsibility are being curtailed such that writers like Elizabeth Ostrom (2000) now refer to the 'crowding out of citizenship'. Thus political involvement is being circumscribed in part by ideologies of self-interest which help to undermine the social conditions of engagement. Conversely, Kymlicka & Norman (1995) suggest that the apparent political passivity of modern citizens may not result from the impoverishment of public life, but rather from the fact that private life is proportionately more enriched than previously and therefore a more attractive focus for one's energies.

Whatever the cause, increasing passivity is agreed, and accompanying it come threats to democracy and constriction of the political space (Altvater 1999) as democratic politics within nation-states loses its legitimacy (Habermas 1976; Beck 1992). Altvater (1999: 48–51) argues that the demands of competing in a global market economy constrain states to align their political logic with the economic logic of global capital, so that we have the irony of growth in the number of formal democratic systems internationally but within states the effect of deregulation and privatisation policies is to diminish the role of the democratic process in social and economic decision-making. The result is a thinning of the substance of democracy and an undermining of the prerequisites of citizenship, including mutual loyalty and trust, a sense of obligation, the sense of having a stake in economic growth, a sense of personal worth, and institutions sufficiently durable to provide a sense of the controllability of events (Sennett 1997; see also Turner 2001 on the effect of a decline in social capital on participatory democracy).

While it may appear that the decline of state power threatens democracy and therefore citizenship, Held (2000) prefers to speak in terms of the transformation of state power and the transformation of democracy. It is a transformation that is in its infancy for although political communities, once exclusively aligned with the nation-state, are being reconfigured through greater global interconnectedness and are therefore tending to overlap (Held 1999), the institutions of democratic accountability that would legitimise the operations of the international system that has arisen to govern transnational relations are yet to be developed (Newby 1996). As Newby (1996: 214) observes:

> The international agencies that comprise this global system not only lack legitimacy but respond only slowly and imperfectly to the changing demands of a disenfranchised and fragmented citizenry . . . Global governance of global resources will require new institutions of democratic accountability if the rights of all global citizens are not to be infringed.

Just as political rights are being curtailed, so are the *social and economic* rights that were won through the labour struggles that culminated in the postwar social welfare consensus. These rights guaranteeing a level of economic welfare sufficient for participation in the social life of society are being progressively whittled away as the socio-economic framework within which they could be enjoyed has been eroded by economic changes, technological innovation and globalisation (Turner 2001). Consequently, the locus of the rights discourse has shifted from the nation-state to the global state with the emergence of three types of *post-national citizenship*, environmental, aboriginal and cultural, connected to three new classes of rights: 'to a safe environment, to aboriginal culture and land, and to ethnic identity', all 'underpinned by a generic right, namely a right to ontological security' (Turner 2001: 206). In being grounded in humans' vulnerability and social precariousness, post-national citizenship essentially transcends national citizenship to reflect the new realities of global risk. This citizenship ideal embodying the human right to ontological security encompasses responsibilities for 'stewardship of the environment, care for the precariousness of human communities, respect for cultural differences and a regard for human dignity' (Turner 2001: 206–7).

Of the additional rights, environmental rights best encapsulate the realignment of the rights discourse from the nation to the globe. This discourse comprises two elements: (1) the rights of nature, and (2) the rights of humans to a healthy environment. On the first account, various theorists have tackled the question of rights for nature, with the majority coming to the view that such rights are hardly feasible (Batty & Gray 1996; Eckersley 1996). However, as Bosselmann (2001) observes, debates around such previously unthinkable rights have helped in the generation of an environmental consciousness.

On the second component of the environmental rights discourse, the right to a healthy environment is taken as a basic human right and is thought to encompass rights to resources as well as a right to liberty (Bosselmann 2001). But a number of reservations have been raised (Thompson 2001). According to Thompson, among the problems with environmental rights is that the ethical basis of rights is too narrow because it is grounded in protecting and achieving individual goods. Consequently, rights are unable to properly account for the motives of individuals, which have proved to be considerably broader than their own well-being. As well, a focus on individual rights inhibits the relations of cooperation necessary to achieve or protect common goods, this being the primary objective of sustainability initiatives.

Global citizenship and civil society?

Recognition of the vulnerability of human existence and its support-ing environment has resulted in proposals for a different, more com-passionate politics that embodies ideological and political alternatives that counteract economic globalisation (Falk 2000). For Held and col-leagues (1999), the challenge is to refashion the democratic political project on a transnational scale and to ensure that risks are shared across the globe. Such a project would entail 'negotiation of a global social contract with market forces that would include environmen-tal protection as a vital element' (Falk 2001: 223). For Falk (2000) and others the future of citizenship resides in a global civil society of which a primary rationale would be the protection of the collec-tive goods of the global commons by seeking to counteract the hyper-individualist trends of recent decades (Wiseman 1998). Such trends, driven as they are by ideologies that foster consumption in the face of clear limits to such growth, appear inimical to aspirations for global sustainability.

Nevertheless, despite the consensus that there is a need for some kind of transnational citizenship and despite evidence of rudimentary intimations of such, questions remain about the means of effective participation (Newby 1996; Macnaghten & Urry 1998; Falk 2000), the ground of identity (Kymlicka 1999), and accountability of global eco-nomic and political forces (Altvater 1999) (see Linklater 2003 for a comprehensive account of the arguments of supporters and opponents of cosmopolitan citizenship). In reality, global civil society is presently characterised more by 'professionalized non-state activism [than by a] cosmopolitan public space' and national governments have been reluc-tant to accept the responsibilities of stewardship that flow from the var-ious conventions and agreements on environment and development (Elliott 2002: 19; see also Macnaghten & Urry 1998 on 'globalisation-from-above'). As well, Beck (2002) has identified three threats to cos-mopolitan societies in (1) a renewed nationalism with ethnic identities and divisions becoming stronger, (2) a free-market ideology that under-mines democratic politics, and (3) the growth of democratic authori-tarianism where the state's capacity to act in an authoritarian manner is increased by force, law and innovations in information technology. In recognition of these realities, Macnaghten & Urry (1998) conclude that any cosmopolitan order is likely to be one of resistance although they suggest that, precisely because it is heterogeneous and cosmopoli-tan, it affords other possibilities.

While there is considerable agreement about the subject of cosmopolitan citizenship, there is also some scepticism. Will Kymlicka (1999), for example, is doubtful that any meaningful form of transnational citizenship can be produced or that transnational institutions can be democratised. The primary obstacle is language. He reasons that meaningful democratic communication can only be carried on and consent achieved in a common language. Although the elites of the European Union may be able to converse comfortably, the reality is that most citizens are confined to political debates in their own vernacular. He concludes: 'language-demarcated political communities remain the primary forum for participatory democratic debates, and for the democratic legitimation of other levels of government' (Kymlicka 1999: 122).

How should we live?: citizenship and sustainability

In the context of sustainability, the key normative question is: 'How should we live?' (Davidson 2000). This is a question that implicitly involves how we relate to other humans and to non-human entities. According to Beck (2002), this question can only now be known globally; it cannot be known locally or nationally. World or cosmopolitan citizenship has grown in significance and possibility as the sustainability discourse has gathered momentum. In the context of global environmental threats, one particular stream of scholarship has transcended cosmopolitan citizenship's somewhat narrow focus on universal human rights – albeit including environmental rights – to encompass responsibilities owed to both non-nationals and non-humans. This is the concept of the *global ecological or planetary citizen* (Thompson 2001) and it is this concept that is proposed to provide the normative content required for a post-national citizenship (Falk 2000).

At the core of a cosmopolitan ethic is the idea of people taking more *responsibility* for the conditions of other people's lives (Elliott 2002). Cosmopolitan citizenship is premised on cooperative relationships that enable the sharing of responsibilities and burdens (Thompson 2001). Like it or not, this era of global risk and vulnerability makes fellow citizens of people across the globe because of their shared responsibilities to participate in the achievement of collective goods such as environmental protection and equity. Citizenship for sustainability entails responsibilities not just for those goods necessary for immediate survival but also for those collective goods that enable the flourishing of humans and other species now and into the future (Davidson 2000). These include the protection of critical ecosystemic services such as those provided by the atmosphere and the oceans, and protection of biodiversity, natural habitats, and cultural heritage. Other virtues appropriate to a sustainability ethic – such as self-restraint, respect and care – have been identified by Barry (1999), Christoff (1996), and Reed & Slaymaker (1993), for example, but in this discussion attention is focused on responsibility to illustrate how apparently similar norms can be deployed for either emancipatory purposes or to yoke citizens to seemingly self-destructive ideologies.

As noted, the normative content of cosmopolitan or global citizenship is focused on responsibility for the human condition. Similarly, embedded within the concept of sustainability in recognition of humans' impact on the health of the planet and of their ecological interdependence is the notion of a shared responsibility to protect and achieve collective environmental goods. For Heater (1999: 137), this amounts to a 'moral consciousness' that can justifiably use the language of citizenship, 'for citizenship entails the ethical element of responsible behaviour towards and obligations to one's fellow citizens and the state or community which provides us with citizenly status'.

Having established that there is an ethical connection between citizenship and sustainability, what remains to be determined is why citizenship is important to sustainability. First, because it is a normative principle, sustainability requires deliberation through a public process of discourse and debate to gain the consent of those whose everyday lives will be affected by this principle and to decide how it should be expressed in law and social policies (Barry 1996). Second, acceptance of global interdependence, where the activities of each individual have widespread ramifications for the lives of others across the globe, has profound implications for the practice of citizenship and its role in the implementation of sustainability. In this respect, changes in individual preferences – lifestyles, practices and values – are a central concern for sustainability because current consumerist lifestyles enjoyed by most Western countries and elites elsewhere appear to be incompatible with norms of global intergenerational and intragenerational equity.

Such an inconsistency is reinforced by an assemblage of precepts relating to unfettered free trade, unencumbered market forces and economic growth – the 'neoliberal ascendancy' (Falk 2001: 232). Conflicts with the normative dimensions of global environmental governance as outlined in the various global conventions on environment and development are therefore quite clear. The source of such variance resides at least partly in the norms of neo-liberal active citizenship that helps to reinforce such ideologies.

Like the concept of the ecological citizen, the neo-liberal understanding of active citizenship presupposes responsible civic practice.

In the latter's case, this involves an obligation on individuals to make themselves 'work-ready' and economically self-reliant (see Kymlicka & Norman 1995 on the New Right's concept of active citizenship and critics' countering arguments). To gain a share in economic success, the good citizen is expected to contribute to economic prosperity by enhancing and promoting their own skills and entrepreneurship, as well as by being a committed consumer. All activity is calculated in terms of investment in self and in family, while activity directed towards the provision of collective or public goods is limited. Critics say that this preoccupation is clearly unsustainable because sustainability largely concerns the provision of public goods. It is also unsustainable in terms of both the limits to throughput of resources and the social division and alienation that are produced. The creation of consumer citizens results in those with the skills, competencies and resources to be good economic citizens being considered 'civilised', while those who lack these skills and who are prevented from participating in economic suc- cess become the 'uncivilised', belonging to a marginalised underclass (Dahrendorf 1994). Unlike the citizenship of entitlement, where stand- ing is independent of economic contribution, for the *obligate* citizen, 'citizenship becomes conditional on conduct' (Rose 2000: 1407).

By contrast, the defining virtue of the ecological citizen is responsi- bility for the condition of fellow humans and the natural world. Here the practice of citizenship becomes one of stewardship (Barry 1999). The importance of the citizen's responsibility for sustainability can best be illustrated by reference to the sense of responsibility associated with the different types of global citizens identified by writers such as Falk (1994), Urry (1999) and van Steenbergen (1994b). Global citizens range along a continuum from the global capitalist or elitist at one pole to the earth citizen at the other (Falk 1994; van Steenbergen 1994b: 147–51). The *global elitist* is characterised as a product of economic glob- alisation, as footloose with little or no attachment to place or commu- nity, and lacking a global civic sense of responsibility. At the opposite pole is the *earth citizen*, who is 'aware of his or her organic process of birth and growth out of the earth as a living organism' and who per- ceives the earth as human habitat (van Steenbergen 1994b: 150). This is an all-inclusive category of citizenship with responsibility extending to the natural world.

The importance of sustainability to citizenship becomes apparent when the possibilities for participation available to the consumer citizen are considered. This class of citizen is a product of the present era of deregulation, wherein many formerly political decisions have been privatised and therefore lack political legitimation (Beck 1992). The result is that economic power overrides political power and citizens can only participate as economic subjects, particularly as consumers (Altvater 1999). Increasingly, avenues for active citizenship are confined to or constructed as consumer protest (Parker 1999). As Parker demonstrates, consumer citizenship can shift government and corporate policy, as occurred in the case of motorway construction in Britain and Shell's Brent Spar oilrig. But he also concludes that, although such action may become an increasingly important tactic of resistance, its effectiveness in achieving systemic change on its own is likely to be limited. As an incremental process, it has only limited efficacy in protecting 'unique or fragile environmental goods' because it can only be a rearguard action (Parker 1999: 79).

As discussed above, citizenship for sustainability requires the active participation of citizens in the decisions that affect their everyday lives. The motivation for participation for sustainability derives from a sense of responsibility not just to fellow citizens but also to the natural world. Participation in this sense 'refers to "being part of" as well as being active in and fully responsible for' (Zweers, cited in van Steenbergen 1994b: 147) the earth as home and habitat. Yet, given the hegemony of contemporary political and economic ideologies, ecological citizenship may appear to belong in the category of utopian ideal. Notwithstanding the many obstacles to participation resulting from contemporary lifestyles and past histories (Davidson 2000), and granted that the traditional means of active participation in society – through work, war, and parenthood (Turner 2001) – are being eroded, Parker's (1999) work on protest citizenship does suggest that alternative means of political participation are opening up, albeit with limited effectiveness.

On the topic of prospects, various contradictory and supporting views have been articulated. Altvater (1999: 57), for example, thinks that the dissonance between political and economic spaces creates spaces for other actors, non-government organisations and civil society, and for changes in the substance of participation. A contrary view is expressed by Howard Newby (1996) that, although there are opportunities for the growth of environmental pressure groups, the conditions that would enable effective participation appear to be largely absent.

FUTURE DIRECTIONS

Discussions around citizenship elaborated in this chapter could very well lead to the conclusion that the odds for a more 'compassionate' politics and therefore a more responsible citizenship are quite small, that the hegemony of the global market economy is almost complete, that there is no room for participatory democracy, and that the chances for the formation of an institutional framework supportive of such a politics on a global scale are negligible. At this stage in what is apparently a period of substantial change in the global order, one can only look to what are called 'edge effects' or 'ecotones' in ecology, for these are the zones of intersecting differences that are pregnant with evolutionary potential (Coles 1992). Two such zones with prospects for further work might include those areas of citizens' resistance to the forces of globalisation, and identification of areas of dissonance and heterogeneity, such as interactions between social movements and governments.

The most pressing issues for the achievement of global environmental security, though, are those related to (1) devising a global social contract that embodies a more equitable sharing of global risks and the benefits of wealth creation; (2) the development of an institutional framework for global governance that ensures the accountability of global corporations; (3) the replacement of consumption ideologies with values more suited to resource limits and global equity; and (4) the development of institutions that promote a balance of rights and responsibilities and foster virtues appropriate to a sustainability ethic. Underlying the first two of these challenges are questions of power – 'the power to define problems, the power to allocate resources and harms, and the power to fashion solutions' (Lutzenhiser 2002: 8). Central to the second two challenges is the normative content of citizenship. For these reasons, the nexus between citizenship and sustainability cannot be theorised apart from issues of power on the one hand and the normative dimensions of citizenship on the other, suggesting that the potential for future fruitful collaboration between sociologists and environmental philosophers is considerable.

Discussion Questions

1. Discuss the prospects for a global civil society and for transnational citizenship.
2. It is argued that the nation-state can no longer act as the ground of identity for citizenship. Place has been suggested as a suitable ground to replace the nation-state as exemplified by the following quote: 'the renewed value on place ... might in fact present an opportunity – the opportunity to construct a public realm in which people think about themselves and act socially other than as economic animals, their value as citizens not dependent upon their riches' (Sennett 1997: 163). Discuss the advantages and disadvantages of this proposal.

3. What are the prospects for international institutions, similar to the Bretton Woods agreement, to control transnational capital movements and to ensure accountability from global economic entities? What form(s) might such institutions of international governance take?

4. Discuss the contention that citizens who fail to contribute to economic success should be stripped of their citizenship rights.

5. How might a spatially layered account of citizenship work in practice? What rights and responsibilities might citizens bear at each level?

6. What kinds of institutions would promote or enforce the virtues and responsibilities of good citizenship? Discuss the essential criteria for the formulation and operation of such institutions.

7. Discuss what policies could promote a balance of rights and responsibilities.

8. Participation is an essential criterion of active citizenship for sustainability. What forms might participation take at different spatial scales – global, national, regional and local?

9. Explain the place of stewardship in an ethic of sustainability and as a norm of ecological citizenship.

10. In numerous international agreements on environment and development, global inequities in the distribution of wealth and risks have been recognised as unsustainable. A global social contract that includes environmental protection has been proposed as a means of sharing both global risks and the benefits of wealth creation. Discuss what such a contract might entail and the prospects for its achievement.

Glossary of Terms

Active citizenship implies active participation in political decision-making beyond the passive citizenship of liberal democracy, which is generally taken to be limited to voting in elections, obeying the law and paying one's taxes.

Consumer citizen: a product of the present era of global change, deregulation and privatisation, which have restricted opportunities for participation in decision-making. The consumer citizen is expected to largely confine his/her active participation to the economic sphere in order to boost the economic success of the nation or region.

Cosmopolitanism: a concept of world order that has its earliest beginnings in the Greek polis and that in modern times was resurrected by Immanuel Kant to negate exclusion among nation-states and to demonstrate that states had strong moral obligations to each other. In recent times the idea of belonging to a wider moral community than that of the nation-state has again been revived in response to the inability to address pressing global problems, such as poverty, inequity, hazards and risks, and environmental degradation.

Economic logic: the term used to describe the suite of ideologies and practices that characterise the phenomenon of economic globalisation, including the primacy of economic growth, free trade, the free market, deregulation, privatisation and managerialist organisation.

Globality: the term that describes a period of human history when the global becomes the primary referent for most human activity.

Globalisation-from-above: a term used by Beck (1992) to describe the dominance of transnational forces, including transnational corporations, international development organisations, and non-government organisations, in the global organisation of human affairs.

Intergenerational equity encapsulates the idea that present generations have obligations to future generations of humans and non-humans.

Neo-liberalism: an economic philosophy rejuvenated from an earlier pure form of economic liberalism that called for no government intervention in economic activities, no restrictions on manufacturing or commerce, and no trade barriers such as tariffs. The main points of this ideology are: (1) unregulated markets, (2) cutbacks in welfare spending, (3) deregulation, (4) privatisation, and (5) replacement of the 'public good' with 'individual responsibility'.

Obligate citizenship: a term that describes neo-liberal or Third Way citizenship where citizens are obliged to ready themselves for the labour market in order to participate in the achievement of the state's/region's economic success; it may also refer to citizens who are obliged to perform 'workfare' or 'work-for-the-dole' activities in return for welfare payments.

Ontological security: the right to ontological security refers to the right to a secure existence; it has only recently emerged as an issue in the context of global problems for which accountability cannot be apportioned and which appear insurmountable.

Political logic refers to the suite of concepts and practices that comprise the liberal-democratic nation-state.

Post-national citizenship describes a form of citizenship where citizens' rights and responsibilities are no longer completely tied to the nation-state because the nation-state is no longer in a position to guarantee those rights, while events have expanded many civic responsibilities from the state to the globe.

References

Albrow, M. 1996, *The Global Age: State and society beyond modernity*, Cambridge: Polity Press.

Altvater, E. 1999, 'The democratic order, economic globalization, and ecological restrictions: on the relation of material and formal democracy'. In Schapiro and Hacker-Cordón, *Democracy's Edges*, pp. 41–62.

Barry, J. 1996, 'Sustainability, political judgement and citizenship: connecting green politics and democracy'. In Doherty and de Geus, *Democracy and Green Political Thought*, pp. 115–31.

Barry, J. 1999, *Rethinking Green Politics: Nature, virtue and progress*, London: Sage.

Batty, H., and T. Gray 1996, 'Environmental Rights and National Sovereignty'. In S. Caney, D. George and P. Jones (eds) *National Rights, International Obligations*, Boulder CO: Westview Press, pp. 149–65.

Beck, U. 1992, *Risk Society: Towards a new modernity*, London: Sage.

Beck, U. 2002, 'The cosmopolitan society and its enemies', *Theory, Culture and Society* 19(1–2): 17–44.

Bosselmann, K. 2001, 'Human rights and the environment: redefining fundamental principles'. In Gleeson and Low, *Governing for the Environment*, pp. 118–34.

Christoff, P. 1996, 'Ecological citizens and ecologically guided democracy'. In Doherty and de Geus, *Democracy and Green Political Thought*, pp. 151–69.

Civic Experts Group 1994, *'Whereas the People': Civics and Citizenship Education*. Report of the Civic Experts Group, Canberra: AGPS.

Cohen, J.L. 1999, 'Changing paradigms of citizenship', *International Sociology* 14(3): 245–68.

Coles, R. 1992, *Self/Power/Other: Political Theory and Dialogical Ethics*, Ithaca NY/London: Cornell University Press.

Dahrendorf, R. 1994, 'The Changing quality of citizenship'. In van Steenbergen, *The Condition of Citizenship*, pp. 10–19.

Davidson, J. 2000, 'Citizenship and sustainability in dependent island communities: the case of the Huon Valley region in southern Tasmania', *Local Environment* 8(5): 527–40.

Dean, H. 2001, 'Green citizenship', *Social Policy and Administration* 35(5): 490–505.

Doherty, B., and M. de Geus (eds) 1996, *Democracy and Green Political Thought: Sustainability, rights and citizenship*, London: Routledge.

Eckersley, R. 1996, 'Liberal democracy and the rights of nature: the struggle for inclusion'. In F. Mathews (ed.) *Ecology and Democracy*, London: Frank Cass, pp. 169–98.

Elliott, L. 2002, 'Global environmental (in)equity and the cosmopolitan project'. CSGR Working Paper, 95/02, ANU, Canberra.

Falk, R. 1994, 'The making of global citizenship'. In van Steenbergen, *The Condition of Citizenship*, pp. 127–40.

Falk, R. 2000, 'The decline of citizenship in an era of globalization', *Citizenship Studies* 4(1): 5–17.

Falk, R. 2001, 'Humane governance and the environment: overcoming neoliberalism'. In Gleeson and Low, *Governing for the Environment*, pp. 221–36.

Gleeson, B., and N. Low (eds) 2001, *Governing for the Environment: Global Problems, Ethics and Democracy*, Basingstoke, UK: Palgrave.

Habermas, J. 1976, *Legitimation Crisis*, London: Heinemann.

Heater, D. 1999, *What is Citizenship?* Cambridge: Polity Press

Held, D. 1999, 'The transformation of political community: rethinking democracy in the context of globalization'. In Schapiro and Hacker-Cordón, *Democracy's Edges*, pp. 84–111.

Held, D. 2000, 'Regulating globalization? The reinvention of politics', *International Sociology* 15(2): 394–408.

Held, D., A. McGrew, D. Goldblatt and J. Perraton 1999, 'Globalization', *Global Governance* 5(4): 483–96.

Jacobs, M. 1995, Reflections on the Discourse and Politics of Sustainable Development Part 1: Faultlines of Contestation and the Radical Model (20pp.), Centre for the Study of Environmental Change, Lancaster University.

Kymlicka, W. 1999, 'Citizenship in an era of globalization: commentary on Held'. In Schapiro and Hacker-Cordón, *Democracy's Edges*, pp. 112–26.

Kymlicka, W., and R. Norman 1995, 'Return of the citizen: a survey of recent work on citizenship theory'. In R. Beiner (ed.) *Theorizing Citizenship*, Albany NY: State University of New York Press, pp. 283–322.

Linklater, A. 2003, 'Cosmopolitan citizenship'. In E.F. Isin and B.S. Turner (eds), *Handbook of Citizenship Studies*, London: Sage, pp. 317–32.

Lutzenhiser, L. 2002, 'Environmental sociology: the very idea', *Organization and Environment* 15(1): 5–9.

Macnaghten, P., and J. Urry 1998, *Contested Natures*, London: Sage.

Marshall, T.H. 1950, *Citizenship and Social Class And Other Essays*, Cambridge University Press.

Newby, H. 1996, 'Citizenship in a green world: global commons and human stewardship'. In M. Bulmer and A.M. Rees (eds) *Citizenship Today: The Contemporary Relevance of T.H. Marshall*, London: UCL Press, pp. 209–21.

Ostrom, E. 2000, 'Crowding out citizenship', *Scandinavian Political Studies* 23(1): 3–16.

Parker, G. 1999, 'The role of the consumer-citizen in environmental protest in the 1990s', *Space and Polity* 3(1): 67–83.

Reed, M.G., and O. Slaymaker 1993, 'Ethics and sustainability: a preliminary perspective', *Environment and Planning A* 25: 723–39.

Rose, N. 2000, 'Community, citizenship, and the third way', *American Behavioral Scientist* 43(9): 1395–411.

Sagoff, M. 1988, *The Economy of the Earth*, Cambridge University Press.

Schapiro, I., and C. Hacker-Cordón (eds) 1999, *Democracy's Edges*, Cambridge University Press.

Sennett, R. 1997, 'The new capitalism', *Social Research* 64(2): 161–80.

Soares, C. 2003, 'Corporate versus individual moral responsibility', *Journal of Business Ethics* 46(2): 143–50.

Spangenberg, J. 2001, 'Towards sustainability'. In Gleeson and Low, *Governing for the Environment*, pp. 29–43.

Thompson, J. 2001, 'Planetary citizenship: the definition and defence of an ideal'. In Gleeson and Low, *Governing for the Environment*, pp. 135–46.

Turner, B.S. 2001, 'The erosion of citizenship', *British Journal of Sociology* 52(2): 189–209.

Urry, J. 1999, 'Globalization and citizenship', *Journal of World-Systems Research* 5(2): 311–24

van Gunsteren, H. 1994, 'Four conceptions of citizenship'. In van Steenbergen, *The Condition of Citizenship*, pp. 36–48.

van Steenbergen, B. (ed.) 1994a, *The Condition of Citizenship*, London: Sage.

van Steenbergen, B. 1994b, 'Towards a global ecological citizen'. In van Steenbergen, *The Condition of Citizenship*, pp. 141–52.

Wallerstein, I. 2000, 'Globalization or the age of transition? A long-term view of the trajectory of the world-system', *International Sociology* 15(2): 249–65.

Wiseman, J. 1998, *Global Nation? Australia and the Politics of Globalisation*, Cambridge University Press.

CHAPTER ELEVEN

THE ENVIRONMENT MOVEMENT
Where to from Here?

Bruce Tranter

In recent decades political elites and the mass public in advanced industrialised democracies have taken on an increasingly green tinge. 'Environmentalists' of one sort or another are to be found just about everywhere, with claims of environmental credibility emanating from the
most unlikely sources. Citizens recycle, politicians issue obligatory policy statements on 'the environment', and protestors take to the streets
or chain themselves to trees in support of urgent environmental concerns. Compared to the middle of the 20th century, most people living
in affluent countries could now be described as 'environmentalists' in
some sense.

BACKGROUND TO THE ISSUES

Organised campaigns for environmental protection began in Australia
in the 19th century (Papadakis 1993), as they did in Western Europe
(Dalton 1994). Notable among the very early Australian campaigns
were those aimed at establishing national parks for recreational
purposes (Papadakis 1993: 66; see also Hutton & Connors 1999).
Environmental issues became particularly salient in the 1960s in Europe
and the United States with the emergence of a 'new wave' of environmentalism (Dalton 1994). Warnings of environmental crises were
heralded in a range of influential books, such as Rachel Carson's *Silent
Spring* (1962) and Paul Ehrlich's *Population Bomb* (1971). These works
focused public attention on a range of environmental 'risks' (Beck
1992), in particular air and water pollution. The overlap with the
student movement was also an important factor in the emergence of
the environmental movement, with university graduates in the 1960s
and 1970s forming 'a leadership cadre and activist core for many of the
newly forming environmental groups' (Dalton 1994: 37).

A surge in support for the Australian environmental movement in the early 1970s followed the failed attempt to save Lake Pedder from flooding by the Hydro-Electric Commission in south-west Tasmania. Although unsuccessful (the lake was flooded in 1972), the campaign to save Pedder saw the formation of the world's first green political party: the United Tasmania Group in 1971 (Hutton & Connors 1999: 121). Activists developed protest and media skills at Lake Pedder, and in other actions such as the 1979 forest campaign at Terenia Creek in New South Wales, that were later employed in a range of successful campaigns. Several activists from the Pedder campaign took leading roles in the Franklin Dam campaign in south-west Tasmania, where a series of protest actions and blockades led to increased public awareness, and put political pressure on the Federal Government to act. This particular campaign resulted in the enactment of legislation by the new Labor Government in 1983 that 'saved' the Franklin River (the authority of the Federal Government to intervene directly in the dispute was subsequently upheld by a High Court ruling to that effect).

The tactics favoured by environmentalists included not only the direct lobbying of political elites and the presentation of petitions (used at Pedder and elsewhere), but the emergence of non-violent direct protest actions, such as blockades (Doyle 2001: 129). Such tactics were employed widely in subsequent campaigns to attract the mass media and thereby influence political elites through the manipulation of public opinion.

Most contemporary Australians are in some sense 'green'. For example, a series of academic surveys conducted throughout the 1990s showed that over 70 per cent of Australians approved of environmental groups (Australian Election Studies). Similarly, a national opinion poll of Australians conducted in 2000 found that 55 per cent agreed with the statement 'I'm a bit of a "Greenie" at heart', while only 26 per cent disagreed (Roy Morgan Research 2000). Support for green politicians is also currently at high levels following the 2001 federal election (Newspoll 2003). In that election the disaffected 'Left' of the Australian Labor Party and also educated cosmopolitans deserted to the Australian Greens in the wake of Labor's prevarication over refugees and asylum-seekers. In corporate Australia, many businesses now report a triple bottom line of not only their economic performance, but also the social and environmental record of companies. In fact some social scientists argue that 'the environment' has become part of the political mainstream or 'routinised' (Crook & Pakulski 1995). If this is the case, what are the implications for the environment movement in Australia, and

elsewhere? The movement has had its share of successes, but will this
very success also contribute to its demise?

KEY DEBATES

What is the environmental movement?

We begin by considering the form and structure of environment movements, the participants involved, and the factors that have made environmentalists influential in advanced industrialised countries. Social
movements have been described by Pakulski (1991: xiv) as 'recurrent
patterns of collective activities which are partially institutionalised,
value oriented and anti systemic in their form and symbolism'. The
environment movement is polymorphous in structure, composed of
networks of individuals, small localised groups or cells, larger environmental movement organisations, and green political parties. While
environmental groups are an important part of the broader movement, a distinction is often made between ecologist and conservation
groups (Lowe & Goyder 1983; Dalton 1994). As Dalton and colleagues
(2003: 758) explain, ecologist groups 'are more likely to focus on the
environmental issues of advanced industrial societies that may call for
basic changes in societal and political relations to address these problems'. Alternatively, conservation groups 'are concerned with wildlife
and other preservation issues and often emphasise these goals without
challenging the dominant social paradigm'. Different types of environmental groups also express their concerns in different ways, with ecologist groups more likely than conservation groups to 'pursue protest
activities . . . and mobilizing activities' (Dalton et al. 2003: 758). Tactics employed by environmental groups vary 'from spectacular forms of
direct action, as in the case of Greenpeace' to 'expert and patient lobbying, the preferred tactics of Friends of the Earth' (Dobson 1990: 3).

Social movement theorists often distinguish 'old and 'new' social
movements (e.g. Offe 1985; Dalton et al. 1990). 'New' social movements (NSMs) such as the environmental, women's and anti-nuclear
movements differ from 'old' social movements such as the labour,
agrarian, and civil rights movements in ideology, support bases, organisational structure and political style (Dalton et al. 1990). In contrast to
old social movements, new ones 'generally advocate greater opportunities to participate in the decisions affecting one's life, whether through
methods of direct democracy, or increased reliance on self-help groups
and cooperative styles of social organisation' (Dalton et al. 1990: 11).
Alternatively, while old movements, particularly the labour movement,

often have a distinct class base, class accounts are of limited use for explaining support for NSMs (Tranter 1996; Mertig & Dunlap 2001).

Social movements tend to be non-hierarchical, anti-bureaucratic and non-institutional in their form (Dalton et al. 2003). As Doyle & McEachern (2001: 55) contend, movements are 'characterised by their informal modes of organisation; their attachment to changing values as a central part of their political challenge; their commitment to open and ultra-democratic, participating modes of organisation . . . and their willingness to engage in direct action to stop outcomes that they see as harmful'. Dalton and colleagues (1990: 5) claim that new social movements 'advocate a new social paradigm which contrasts with the dominant goal structure of Western industrial societies'. Rootes (1999: 2) highlights the diversity of the environmental movement with 'organisational forms ranging from the highly organised and formally institutionalised to the radically informal'.

As should be apparent, it can be difficult to pin down what social movements actually consist of because of their fluid form and loose structure, although they do have organisational cores. The Wilderness Society, the Australian Conservation Foundation, Greenpeace, and Friends of the Earth are examples of key environmental movement organisations (EMOs). EMOs have members and/or financial contributors, with some organisations having large memberships. Members and participants are linked through informal networks of volunteers who provide these organisations with the resource base and people power for potential mobilisation. EMOs are also focal points for mass media. Journalists use them as contact points, or more typically approach particular individuals within EMOs who act as environmental spokespeople (Tranter 1995).

However, while environmental movement organisations are key players in the articulation of social movement concerns, it should be stressed that the environmental movement per se consists of much more than just high-profile organisations and green political parties. Social movements are loose collectivities comprising networks of individuals, local groups and organisations (Doyle & McEachern 2001). Some environmentalists involved in social movements are members of EMOs, but others belong only to informal, localised groups, or grassroots cells. In the environment movement, small cells often form rapidly in response to local environmental threats but may also disappear quickly when the threat has passed. EMOs serve as focal points, not only for mass media, but also for disparate groups of potential activists who lack organisational experience and are not skilled in protest actions. As Crook

Table 11.1 *Membership of environmental groups, 1990–2001 (%)*

	1990	1993	1996	1998	2001
How likely are you to join an environmental group or movement? Already a member	2.9	4.5	2.4	n/a	5.4
Strongly approve or approve of environmental groups?	73.3	73.2	71.3	76.3	n/a
	(2037)	(2388)	(1797)	(1897)	(2010)

Source: Australian Election Studies (1990–2001)

and Pakulski (1995: 40) put it, 'green movements and parties publicise issues by making them the focus of their protest activities and electoral appeals; these appeals, in turn, help to amplify the issues and organise them into relatively coherent sets of public orientations and concerns.'

Support for environmental groups

How much support do EMOs receive? Table 11.1 shows the proportion of Australian adults who have joined environmental groups or approve of environmental groups. Results from the nationally representative Australian Election Study survey (Bean et al. 2002) show that more than 5 per cent of Australians claimed to be members of an environmental group in 2001, with public approval for these groups running at more than 70 per cent during the last decade. However, as we are unable to disaggregate responses into more 'activist' groups such as The Wilderness Society, and 'passive' groups such as Landcare, these responses provide only a rough guide to support for environmental groups.

Environmental activists in Australia have often sought to claim the moral high ground through their choice of issues, not only through the value-laden (Pakulski 1991) causes they pursue (e.g. preservation of wilderness areas, protection of endangered species) but also by engaging in generally non-violent protest actions. Australian activists have tended to avoid violent tactics such as tree or road spiking, or the destruction of machinery referred to in the literature as 'monkey wrenching' (Foreman & Haywood 1987). While there have been some instances of 'monkey wrenching' in Australia, Doyle (2001: 49) points out that 'non-violent action has dominated environmental direct actions during major campaigns'. Non-violent protest actions include 'marches, rallies, pickets, strikes, boycotts, sit-ins and block-ades' (Doyle 2001: 49), and while they are peaceful, they tend to be colourful, noisy and confronting forms of protests. Participants engage in rallies and marches, blockades and tree-sitting, or unfurl banners from high-rise buildings or bridges in order to gather public support by attracting mass media, particularly television, to publicise environmental issues.

Movement leadership

The loose structure of the environment movement has implications for power and leadership within it. In contrast to formal organisations, the movement is polycephalous, and EMOs aside, consists as we have seen of 'loosely connected groups, social circles and networks, as well as a large number of supporters and sympathisers who are not affiliated with any organised bodies' (Pakulski 1991: 43). Doyle (2001: 11) has used the term 'elite' when discussing power within the movement, claiming for example that 'in the wet tropics campaign . . . a small band of professional, organisational activists banded together to dominate many conservation initiatives. As such elites increase their hold on movement politics, representativeness and equality in decision-making diminish.'

In contrast to government, business, trades union and other elites, however, environmental 'leaders' do not have formal authority over their followers. Movements are not merely comprised of EMOs, and even EMOs tend to be non-bureaucratic and non-hierarchical (Pakulski 1991), although the hierarchically organised Greenpeace is an exception (see Burgmann 2003: 198). Participants in environmental movements are usually volunteers, with only small numbers of activists employed by EMOs. Participants in new social movements typically reject non-democratic decision-making processes (Dalton et al. 1990), preferring participatory-style democratic decision-making reached through consensus.

The concept 'elite', as it is usually applied in sociology and political science, implies the existence of hierarchical, organisation-based power, where leaders have legitimate authority over subordinates (Higley et al. 1979). It is therefore not appropriate to apply the term 'elite' to 'leaders' in the broader environmental movement, even if they are within EMOs. Environmental leaders exert a more subtle form of influence over other movement participants, influence that often stems from respect gained within the movement by involvement in substantial and successful campaigns (Tranter 1995). This lack of clear authority has important implications for the direction of the movement, because even if 'leaders' attempt to 'lead', other activists and supporters are not compelled to follow.

Green politicians

How do green politicians fit into the movement? Rootes (1999: 1) argues that greens 'are established players in the political arena of most western European states . . . widely regarded as part of a broader green movement' and 'the fact that greens have entered government in an increasing number of states owes more to the perceived strength and popularity of environmental movements than it does to the generally marginal electoral performances of green parties themselves' (Rootes 1999: 1). Yet the appearance of greens in representative politics is a fairly recent phenomenon in Australia. Electoral success of course depends in part on favourable political opportunity structures (Muller-Rommel 1989). In Australia, green politicians, and for that matter minor parties, have a far better chance of electoral success in the pro-portionally representative Senate elections than under the preferential system in the legislative House of Representatives.

Tasmania is arguably the greenest Australian state, at least in terms of representative politics, with the electoral success of the Greens aided by the proportional Hare–Clarke electoral system. While the success of the Greens has fluctuated since the late 1980s, they currently hold four state seats in Tasmania (out of 25 seats in total; the Labor Party holds 14, and the Liberal Party 7), and nationally have two senators, including the most prominent and electorally successful Australian green politi-cian, Dr Bob Brown. Green politicians in other states have been far less successful at the state and federal level, although at the federal level the Greens achieved a milestone in 2002 with the election of their first member of the House of Representatives (Burgmann 2003).

Key environmental issues

While some environmental issues such as global warming and the greenhouse effect are obviously of global or cross-national concern, support for environmental issues is differentiated to a large degree on a national basis. In Europe and the United States important issues include concern over pollution, acid rain, toxic waste and nuclear power (Dalton 1994). In Australia, McAllister (1994: 22) identified three dimensions of environmentalism: 'a cosmopolitan dimension, encompassing national and international concerns; a local dimension focussing on general concerns; and a local dimension concerned solely with damage to land'. Crook and Pakulski (1995) and Pakulski and

Table 11.2 *Urgency of environmental issues in Australia (%)*

	Adult Australians		Environmental group members	
	Very urgent	Rank	Very urgent	Rank
Pollution	39.5	4	67.9	4
Waste disposal	35.8	6	54.7	6
Logging of forests	38.1	5	71.7	3
Destruction of wildlife	44.8	2	75.5	1
Soil degradation	45.2	1	73.6	2
Greenhouse effect	40.9	3	60.4	5
Genetically modified crops/food	28.5	8	32.4	8
Cloning human tissue	29.6	7	37.1	7

Responses from the question: 'In your opinion, how urgent are each of the following environmental concerns in this country?' Response categories on five point scales ranged from 'not urgent' to 'very urgent' with a neutral midpoint. Percentages indicating 'very urgent' responses are reported.
Source: Australian Election Study survey 2001.

colleagues (1998) classified issues into 'green' and 'brown' clusters. They argued that the 'green' issue-cluster (i.e. logging of forests and wildlife preservation) is most closely associated with the goals of environmental groups, while 'brown' concerns (over pollution and waste disposal) 'have no obvious or natural ideological/discursive home, just as their constituency has no specific cultural, social or political milieu' (Crook & Pakulski 1995: 54). In research that drew on Pakulski and Crook's 'green' and 'brown' typology, McAllister & Studlar (1999: 790) divided members of environmental groups into 'committed' and 'ordinary' members. They found that committed members were 'motivated by a strong sense of the urgency of green environmental concerns, an urgency which is largely absent among ordinary members and the rest of the population'.

The salience of environmental issues can be assessed by again drawing upon the 2001 Australian Election Study data. Table 11.2 shows the proportion of Australian adults who saw particular environmental issues as very urgent, contrasted with the priorities of environmental group members. The most urgent environmental issues for both Australian adults and group members were soil degradation and the destruction of wildlife, although members of environmental groups ranked wildlife preservation as most important, with the public prioritising soil degradation. The 'green' issue, logging of forests, ranked third among group members, closely behind wildlife and soil degradation, but only fifth for all Australians, perhaps reflecting the issue priorities of 'activist' groups such as The Wilderness Society. GM food/crops and the cloning of human tissue – dubbed 'white' issues by Pakulski & Tranter (2003) in reference to the white laboratory coats of scientists – received by far the least support. While 'white' issues have featured prominently in the press and raise important ethical concerns, they are not given priority by the Australian public, and are relatively unimportant even for members of environmental groups.

The social and political bases of the Australian environment movement

Environmentalists are drawn from all walks of life, although environmental supporters tend to have certain social characteristics. Researchers considering the social bases of the Australian environmental movement (e.g. Papadakis 1993; Tranter 1996, 1999; McAllister & Studlar 1999), like those who have examined the situation in Europe and the United States (e.g. Bean 1998; Mertig & Dunlap 2001), found that environmental activists tend to be young and were more likely to be women than men, and that support for environmental concerns is higher in large cities. Australian environmentalists tend to be politically left of centre, non-religious, and highly educated, while education in the humanities may predispose one towards environmental activism (Tranter 1997).

Better known for his research on value change across a range of countries, Inglehart (1977, 1990a, 1990b, 1997) has highlighted the importance of the value priorities held by environmentalists. Inglehart's 'silent revolution' thesis posits a major shift in the value priorities held by citizens of advanced industrialised countries. He has argued that to understand political preferences and behaviour it is necessary to consider the socialisation experiences of successive generations, as early experiences influence the formation of different value priorities.

Growing up under circumstances of relative economic affluence and physical safety leads to the prioritisation of quality of life and post-material issues over economic, materialist issues. Alternatively, experience of economic hardship, war or major social and political upheaval contributes to the development of *materialist* values (Inglehart 1977: 23). Values formed in childhood and early adolescence remain relatively stable over the lifecourse (Inglehart 1997: 34, 46), so that generation-based value differences are apparent across birth cohorts. Younger generations, particularly those born after the Second

World War, tend to prioritise post-materialist values due to childhood experiences of relative affluence and safety. Therefore citizens of advanced Western nations are becoming increasingly post-materialist with generational replacement. Those born after the Second World War tend to hold more favourable opinions on the environment, with post-materialists 'more likely than Materialists to approve of the environmentalist cause and *much* likelier to act on its behalf' (Inglehart 1995: 68).

How well do these claims of environmental support based around a new class, education, generations, and values sit when evaluated using the most recent Australian data? The social bases of environmental support in Australia were assessed using a composite 'environmental new politics' scale that combines support for the environment as an election issue with the likelihood of joining environmental groups. The scale was analysed using the ordinary least squares regression method (Table 11.3).

Net of other influences, women were more supportive of the environmental movement than men, as were the tertiary-educated, professionals, and secular Australians. Concern for the environment among the highly educated and professional lends some support to the new class thesis, although self-identified class location had little impact. Contrary to earlier Australian findings, living in *rural* areas appears conducive to environmental support, perhaps indicating a greening of the bush, or reflecting rural support for preservationist types of environmental EMOs (e.g. Landcare). These findings also mirror research from the United States that suggests the rural/urban divide in environmental support has weakened (Jones et al. 1999: 482). Post-material value orientations and left-wing ideology were strong predictors of environmental new politics (Crook & Pakulski 1995; Tranter 1996). Some post-Second World War age cohort effects were apparent, lending some support to Inglehart's claims of generational-based support, although these effects were relatively weak.

Bifurcation of environmental groups

In recent decades the major environmental issues and concerns for the Australian movement have centred on wilderness conservation and the prevention of forest logging (Pakulski & Crook 1998). Environmental group support for 'green' issues 'suggests the beginnings of a dissociation between the environmental concerns of the environment movement and those of the wider public' (Crook & Pakulski 1995: 53). The bifurcation of issues into 'green' and 'brown' is also apparent in different

Table 11.3 *Social bases of environmental new politics (OLS regression)*

Intercept	29.8
Men	−4.4***
Birth Cohorts	
1966–75	2.4
1956–65	5.4*
1946–55	4.3*
1936–45	2.8
1926–35	4.4
(reference 1900–1925)	
Degree	4.0*
Professional	5.1**
Government Employee	1.7
Large City	1.1
Rural area or village	5.1**
(reference suburbs, small or large towns)	
Secular	5.2***
Middle Class Self Identification	0.9
Postmaterial Values (+ Postmaterial)	11.5***
Political Orientation (+ left)	15.5***
R^2	.11

* p<.05; ** p<.01; *** p<.001.
Dependent variable is a scale constructed from the following questions: 'Here is a list
of important issues that were discussed during the election campaign. When you were
deciding about how to vote, how important was each of these issues to you person-
ally? – The Environment' [Response categories: extremely important (46.7 per cent);
quite important (44.2 per cent); not very important (9.1 per cent)]. 'How likely are you
to join any environmental groups or movements?' [Response categories: I'm already a
member (5.4 per cent); Not a member, but have considered joining (20.8 per cent);
Not a member, and have not considered joining (46.9 per cent); Would never consider
joining (26.9 per cent)]. Scale was rescored to range between 0 (low support) and 100
(high support for the environment).
Source: Australian Election Study survey 2001.

EMOs. Activist-centred EMOs tend to prioritise issues such as the
preservation of old-growth forests (e.g. TWS, Greenpeace), while 'passive' groups (e.g. Landcare, Coastcare) concentrate on re-establishing
vegetation previously native to a particular area, or on addressing problems such as soil erosion and salinity.

Passive environmental groups were the main winners in 1996, when
the Australian Federal Coalition Government set aside $1.5 billion
from the partial sale of the then publicly owned telecommunications
organisation, Telstra, to be spent on 'the environment'. For example,

the Federal Government's Natural Heritage Trust was established in order to help 'communities undertake local projects aimed at conserving biodiversity and promoting sustainable resource use' (Natural Heritage Trust website). Passive groups such as Landcare Australia find healthy support in rural communities, at least in part because they benefit landholders. Similarly, 'Clean up Australia' campaigns attract widespread volunteer support across Australia to remove litter from Australian waterways. Supported by a broad cross-section of people, such groups and campaigns comprise the 'government-initiated' (Doyle 2001: 34), respectable, non-activist face of the environmental movement. Pollution of waterways, waste disposal and soil erosion are broad issues that appeal to a majority of Australians. Supported by government funding, many passive groups present a politically neutral front, and politicians across the ideological spectrum gain credence by supporting 'brown' issues and groups.

'Green' environmental EMOs and the issues they champion, however, are much more politically charged, and participants in these groups are less willing to compromise their principles. Issues prioritised by environmental activists are not supported as strongly by the mass public or by state and federal governments. While the environment movement remains well organised and adept at manipulating the mass media, opposition 'backlash' groups have arisen to fight for sectional interests such as the forest industry (see Rowell 1996), and are increasingly able to influence governments. A potential problem for the more radical activist core of the environment movement arises from their tendency to pressure governments indirectly via the mobilisation of public opinion, to a greater extent than working within the realm of conventional politics. This may lead to their political marginalisation, particularly when conservative governments hold power.

Sceptics could interpret government support for groups such as Landcare as examples of 'greenwashing' – where governments or business organisations present a green façade in order to divert attention or further their own interests. For example, Greer & Bruno (1996: 12) argue that by using 'a sophisticated greenwash strategy' transnational corporations work 'to manipulate the definition of environmentalism and of sustainable development, and to ensure that trade and environment agreements are shaped, if not dictated, by the corporate agenda'. Greer & Bruno (1996) claim that transnational corporations began to adopt greenwash strategies in the late 1980s as the environmental movement gained political prominence, then later adopted the buzzwords 'sustainable development', which abounds in the rhetoric of big business and government. A further example of greenwashing is the financial, environmental and social 'triple bottom line' now popular in the corporate sector (see Donoghue et al. 2003).

FUTURE DIRECTIONS

Routinisation of environmental concerns and its consequences

Australian sociologists have argued that environmental concerns have become routinised (Crook & Pakulski 1995), referring to the claim that environmental issues are no longer new and radical but have become part of the mainstream. Pakulski and Crook (1998: 12) maintain that the process of routinisation 'involves a shift from the new and special to the expected and familiar'. In a related work Pakulski & Tranter (2003: 3) also argued that 'the decline in public concerns followed a change in the way environmental issues were covered by the popular media' as 'coverage of "the environment" changed from sensational to regular'. To an extent this is a normal occurrence as new issues enter the public agenda and compete for attention with established issues (Pakulski & Tranter 2003: 3).

However, the process of routinisation has implications, particularly for 'activist' environmental groups, such as the problem of activist recruitment. New environmental issues have the power to attract new supporters and activists, but the movement in Australia has now reached a point where it is in danger of stagnating. In the United Kingdom, the rise of 'radical' direct actions by groups such as Earth First! and the anti-roads movement occurred at least partly in response to the political conservatism and lack of opportunity for radical protest action in established environmental organisations (Wall 1999; Smith 2000). Wall (1999) argued that a large pool of young activists was created in the United Kingdom by factors such as an expanded higher education system and high unemployment. There are similarities in Australia with its expanding higher education system and youth unemployment, although radical environmentalists are relatively less common. The issues championed by activist-oriented EMOs in Australia, such as the preservation of old-growth forests, continue to attract a small core group of young, educated, but non-violent activists.

As many influential figures in the movement have been involved since the protest actions of the 1980s, the further problem of leadership succession also arises. Young environmentalists may be lured by the importance of environmental causes, but like new recruits in many organisations they often have to undergo a fairly long period of 'apprenticeship' before becoming influential in their own right. This problem is exacerbated by the fact that high-profile activists in EMOs effectively act as movement representatives, or spokespeople, with journalists tending to seek out particular activists and green politicians. High rates of attrition can occur under such circumstances, with few new 'leaders' emerging to replace the ageing cadre of baby boomers who have been influential in the movement for decades.

In reference to earlier Tasmanian protest actions, Montgomery (1998: 88) argued:

> The Franklin was the first peak environmental story to run consistently on page one, but the conduct of the campaign set a very high standard . . . Later campaigns, notably the Lemonthyme and the Tarkine were unable to emulate the Franklin. The sheer size of the operation, the remoteness of the location, the nature of the people who were taking part gave it a legitimacy, which later campaigns have lacked.

Montgomery was obviously sceptical of the environment movement's ability to continue to attract mass media without changing their tactics radically. While this chapter was being written, Wilderness Society and Greenpeace activists were perched on a 65m-high platform in a tree in the Tasmanian Styx valley. Such protests are spectacular, yet their continuing efficacy in attracting mass media coverage is problematic. This is exacerbated when hostile companies or governments restrict access to media organisations, so that the public influence of environmental protest actions is effectively silenced.

Support for the environment varies widely according to the issue. Wild rivers such as the Franklin in Tasmania or rainforests such as the Daintree in Far North Queensland are areas of spectacular natural beauty that appeal to a broad range of people (even if they rarely if ever visit these places) and make for impressive visual representation in a variety of media. Alternatively, while dry forests also have great intrinsic value, they are not so easily represented visually, and therefore may be less appealing to the public. While issues such as forest protection and global warming remain high priorities for environmental activists, mobilising public opinion in response to such issues is problematic for movement activists and leaders.

International terrorism and the environment movement

After the events of 11 September 2001, Australia has been a member of the 'coalition of the willing' in the war against Iraq and as an ally of the United States is engaging in the so-called 'War on Terror'.

Is it likely that the perceived risk of terrorism may have an impact on support for the environment movement? If Inglehart's value shift thesis is correct, there may well be an association between these unlikely bedfellows, although the effect may not be apparent for some years.

As we have seen, Inglehart suggests that those born during periods of relative affluence and physical security tend to develop post-materialist value priorities, while materialist values are likely to develop during times of economic hardship or threats to safety. Post-materialist generations, he argues, are in the ascendancy in advanced industrialised countries, and underpin the support base of new social movements, including the environment movement. However, a shift back towards materialist value priorities in response to changing economic circumstances or the emergence of threats to physical safety is also possible (Inglehart & Abramson 1994: 351).

What impact could the continued threat of terrorism have on the environment movement? If Inglehart's thesis holds, young people growing up in the shadow of international terrorism should tend to form materialist rather than post-materialist value orientations. A value shift back towards materialism should also see a concomitant shift away from support for 'the environment'. The support base of environmental social movements, underpinned mainly by post-materialists, should therefore diminish, particularly among younger citizens, as they will be more concerned with basic safety and security matters than their predecessors. Whether such a scenario will actually occur is an empirical question. It will depend on the efficacy and longevity of terrorist activities, on citizens' perceptions of risk, and, of course, on the validity of Inglehart's value shift thesis. If the threat of terrorist actions does influence the formation of values, however, support for environmental groups and issues may well be attenuated, not only in Australia but also globally.

Discussion Questions

1. What is an environmental movement?
2. Can most people be described as 'environmentalists'? Explain your answer.
3. What impact will the routinisation of environmental issues have on the environment movement?
4. To what extent are more 'radical' environmental activists able to maintain their political influence without compromising over core issues?
5. How can environmental organisations attract new members?
6. How will the ageing of the 'baby boom' generation affect the environment movement?

7. In what ways do leaders in the environment movement differ from leaders in business or political parties?

8. Environmental groups attempt to gather public support by attracting mass media coverage of key issues. How can media interest be maintained as green issues enter the mainstream?

9. What is 'greenwashing'? Why is it of concern to environmental movement activists?

10. Has 'the environment' waned as a political issue, or is it remaining salient since entering the policy agendas of major political parties?

Glossary of Terms

Bifurcation of environmental groups: a divide in Australian environmental groups and organisations between activist-centred, politically radical, protest-oriented groups concerned with 'green' issues, and passive, politically neutral or conservative groups pursuing 'brown' issues.

Environmental issues exist on a variety of levels: international, such as global warming and the greenhouse effect; cross-national and national, such as acid rain and river pollution; and state, such as the prevention of logging or wildlife preservation. Localised issues also precipitate the emergence of new environmental groups.

Environmental leaders tend to have influence over other NSM participants rather than organisation-based authority. Influence and credibility within the movement are gained over time through involvement as activists or as successful campaign organisers. Leadership roles include spokespeople, green politicians, experts, organisers, charismatic figures and elder statespersons.

Environmental movement organisations: organisations oriented to a range of 'green' and/or 'brown' environmental issues. EMOs are often staffed by small numbers of paid employees, but draw on large numbers of part-time volunteers. Participatory democratic decision-making is followed in principle, although high-profile activists tend to set the agenda.

Green politicians: members of green political parties, such as the Australian Greens, or independents with close links to EMOs. They are often influential activists before entering representative politics.

New social movements: loose affiliations of organisations, groups and cells linked through informal networks. Examples include environmental, anti-nuclear, peace and women's movements. NSMs tend to be non-hierarchical and anti-bureaucratic. Participants favour participatory democratic decision-making processes.

Post-material value orientations: Inglehart argues that value formation is influenced by the socio-economic circumstances prevailing during one's formative years. Values differ across generations. Younger people tend to hold post-materialist values and prioritise quality-of-life issues such as the environment. Older generations tend to prioritise materialist, economic issues.

Social and political bases of the environment movement: important social and political characteristics that typify environmental group members and activists. For example, gender, generation, education, social class, location, and political orientation (left/right).

References

Bean, C., I. McAllister and D. Gow 2002, *Australian Election Study* [computer file], Canberra: Social Sciences Data Archives, ANU.
Bean, C. 1998, 'Australian attitudes towards the environment in cross national perspective'.In Pakulski and Crook, *Ebbing of the Green Tide?*
Beck, U. 1992, *Risk Society: Towards a new modernity*, transl. Mark Ritter, London: Sage.
Burgmann, V. 2003, *Power, Profit and Protest: Australian social movements and globalisation*, Sydney: Allen & Unwin.
Carson, R. 1962, *Silent Spring*, Boston MA: Houghton Mifflin.
Crook, S., and J. Pakulski 1995, 'Shades of green: public opinion on environmental issues in Australia', *Australian Journal of Political Science* 30: 39–55.
Dalton, R. 1994, *The Green Rainbow: Environmental groups in Western Europe*, New Haven CT: Yale University Press.
Dalton, R., M. Kuechler and W. Burklin 1990, 'The challenge of new movements'. In Dalton and Kuechler, *Challenging the Political Order.*
Dalton, R., and M. Kuechler (eds) 1990, *Challenging the Political Order: New social and political movements in Western democracies*, New York: Oxford University Press.
Dalton, R., S. Recchia and R. Rohrschneider 2003, 'The environmental movement and the modes of political action', *Comparative Political Studies* 36(7): 743–71.
Dobson, A. 1990, *Green Political Thought: An introduction*, London: Harper.
Donoghue, J., B. Tranter and R. White 2003, 'Homeownership, shareownership and Coalition policy', *Australian Journal of Political Economy* 52(2): 58–82.
Doyle, Timothy 2001, *Green Power: The environment movement in Australia*, Sydney: UNSW Press.
Doyle, T., and D. McEachern 2001, *Environment and Politics*, London: Routledge.
Ehrlich, P. 1971, *The Population Bomb*, New York: Ballantine Books.
Foreman, D., and B. Haywood 1987, *Ecodefense: A field guide to monkey wrenching*, 2nd edn, Tucson AZ: Ned Ludd Books.
Greer, J., and K. Bruno 1996, *Greenwash: The reality behind corporate environmentalism*, New York: Apex Press.
Higley, J., D. Deacon and D. Smart 1979, *Elites in Australia*, London: Routledge & Kegan Paul.
Hutton, D., and L. Connors 1999, *A History of the Australian Environment Movement*, Cambridge University Press.
Inglehart, R. 1977, *The Silent Revolution: Changing values and political styles among western public*, Princeton University Press.

Inglehart, R. 1990a, *Culture Shift in Advanced Industrial Societies*, Princeton University Press.

Inglehart, R. 1990b, 'Values, ideology and cognitive mobilisation in new social movements'. In Dalton and Kuechler, *Challenging the Political Order*, pp. 43–66.

Inglehart, R. 1995, 'Public support for environmental protection: objective problems and subjective values in 43 societies', *PS: Political Science and Politics* March: 57–71.

Inglehart, R. 1997, *Modernization and Postmodernisation: Cultural, economic and political change in 43 societies*, Princeton University Press.

Inglehart, R., and P. Abramson 1994, 'Economic Security and Value Change', *American Political Science Review* 88: 336–54.

Jones, R., H. Fly and H. Cordell 1999, 'How green is my valley? Tracking rural and urban environmentalism in the Southern Appalachian Ecoregion', *Rural Sociology* 64(3): 182–99.

Landcare website http://www.landcareaustralia.com.au/faq.asp (accessed 16 December 2003).

Lowe, P., and J. Goyder 1983, *Environmental Groups in Politics*, London: Allen & Unwin.

McAllister, I. 1994, 'Dimensions of environmentalism: public opinion, political activism and party support in Australia', *Environmental Politics* 3(1): 22–42.

McAllister, I., and D. Studlar 1999, 'Green versus brown: explaining environmental commitment in Australia', *Social Science Quarterly* 80(4): 775–95.

Mertig, A., and R. Dunlap 2001, 'Environmentalism, new social movements, and the new class; a cross-national investigation', *Rural Sociology* 66(1): 113–36.

Montgomery, B. 1998, 'But where's the blood!'. In Pakulski and Crook, *Ebbing of the Green Tide?*, pp. 85–9.

Muller-Rommel, F. 1989, 'The German Greens in the 1980s: short-term cyclical protest or indicator of transformation', *Political Studies* 37: 114–22.

National Heritage Trust website http://www.nht.gov.au/envirofund/index.html (accessed 16 December 2003).

Newspoll 2003, 'Geographic and demographic analysis: voting intention and leaders' ratings', *The Australian* 22 December 2003. http://www.newspoll.com.au/cgi-bin/display_poll_data.pl.

Offe, C. 1985, 'New social movements: challenging the boundaries of institutional politics', *Social Research* 52(4): 817–67.

Pakulski, J. 1991, *Social Movements: The politics of moral protest*, Melbourne: Longman Cheshire.

Pakulski, J., and S. Crook (eds) 1998, *Ebbing of the Green Tide? Environmentalism, public opinion and the media in Australia*, Hobart: University of Tasmania, School of Sociology and Social Work.

Pakulski, J., and B. Tranter 2003, 'Environmentalism and social differentiation: a paper in memory of Steve Crook', Australian Sociological Association conference, Armidale, University of New England, 4–6 December.

Pakulski, J., B. Tranter and S. Crook 1998, 'Dynamics of environmental issues in Australia: concerns, clusters and carriers', *Australian Journal of Political Science* (33)2: 235–53.

Papadakis, E. 1993, *Politics and the Environment: The Australian experience*, Sydney: Allen & Unwin.

Rootes, C. 1999, 'Environmental movements: from the local to the global', *Environmental Politics: Local, National, Global*, Special Issue 8(1): 1–12.

Rowell, A. 1996, *Green Backlash: Global subversion of the environmental movement*, London: Routledge.

Roy Morgan Research 2000, 'Australians Find It Easy Being Green' (Finding No. 3309), http://oldwww.roymorgan.com/polls/2000/3309/index.html.

Smith, G. 2000, 'Radical activism', *Environmental Politics* 9(4): 150–3.

Tranter, B. 1995, 'Leadership in the Tasmanian environmental movement', *Australian and New Zealand Journal of Sociology* 31(3): 83–93.

Tranter, B. 1996, 'The social bases of environmentalism in Australia', *Australian and New Zealand Journal of Sociology* 32(2): 61–84.

Tranter, B. 1997, 'Environmentalism and education in Australia', *Environmental Politics* 6(2): 123–43.

Tranter, B. 1999, 'Environmentalism in Australia: elites and the public', *Journal of Sociology* 35(3): 331–50.

Wall, D. 1999, *Earth First! and the Anti Roads Movement: Radical environmentalism and comparative social movements*, London: Routledge

CHAPTER TWELVE

MOULDING AND MANIPULATING THE NEWS

Sharon Beder

The media are accused of bias by people from both ends of the political spectrum, but journalists, editors and owners maintain that they pro-vide an objective source of news. This chapter will consider the ways in which the news is shaped and how this in turn influences the way environmental issues are reported and constructed in the mass media.

In the United States, where the debate over media objectivity is most heated, conservatives criticise the media for having a 'liberal' bias and these critics focus on the personal views of journalists, editors and media owners who, they argue, tend to be elitist, left-leaning and politically correct. A number of books have been published recently highlighting this supposed liberal bias including *Press Bias and Politics: How the media frame controversial issues* (Kuypers 2002), *Bias: A CBS Insider Exposes How the Media Distort the News* (Goldberg 2003b) and *Arrogance: Res-cuing America from the media elite* (Goldberg 2003a). Goldberg argues that 'the majority of journalists in big newsrooms slant leftward in their personal politics, especially on issues like abortion, affirmative action, gay rights, and gun control; and so in their professional role they tend to assume those positions are reasonable and morally correct. Bias in the news stems from *that* . . .' (Goldberg 2003a: 4).

Accusations of liberal bias, however, are not new. Richard Nixon and his vice-president Spiro Agnew repeatedly referred to the bias of the media, particularly with regard to the Nixon administration, and Agnew called journalists 'pointy-headed intellectuals' (West 2001: 65). Corporate executives and conservative leaders attributed the surge of regulation and the distrust of business of the late 1960s and early 1970s in part to the media and what they perceived as the media's liberal bias. As part of the political resurgence of conservative ideas they sought to build their own reliable media outlets and to have more influence over existing media organisations.

Robert Parry, author of *Fooling America*, describes a well-financed plan to build a conservative press in the United States: 'It ranges from nationwide radio talkshows . . . to dozens of attack magazines, newspapers, newsletters and right-wing opinion columns, to national cable television networks propagating hard-line conservative values and viewpoints, to documentary producers who specialize in slick character assassination, to mega-buck publishing houses' (Parry 1995: 6).

Most conservative organisations produced their own publications or media programs. Corporate-funded think-tanks and public relations firms recruited journalists from the mainstream media to their own staffs. Conservative student newspapers were financed, as were conservative television programs such as Milton Friedman's series *Free to Choose*, which was broadcast on the Public Broadcasting Service (PBS). So much oil company money went into sponsoring PBS programs that it was nicknamed the Petroleum Broadcasting Service (Saloma 1984: 107).

The 'liberals' themselves accuse the media of a conservative bias but focus more on institutional factors than on individual biases. Recent examples include *What Liberal Media? The truth about bias and the news* (Alterman 2003); *Censored 2001: 25 years of censored news and the top censored stories of the year* (Phillips et al. 2001), whose authors publish the top stories that are *not* reported by the mass media each year, and the latest edition of *Manufacturing Consent: The political economy of the mass media* (Herman & Chomsky 2002). Herman & Chomsky (2002: 2) developed a 'propaganda model' of the mass media in the United States by 'tracing the routes by which money and power are able to filter out the news fit to print, marginalize dissent, and allow the government and dominant private interests to get their messages across to the public'. They argue that news is subject to a number of filters which include media owners and advertisers and news sources.

BACKGROUND TO THE ISSUES

The concentration of media ownership into the hands of a few people is of concern to people at both ends of the political spectrum. For example, Rupert Murdoch controls more than half the newspapers in Australia, including the only major daily newspaper in Brisbane, Adelaide and many regional cities, and is lobbying for changes to Australian media ownership laws to enable him to buy a television network (Lawson 2003).

A pattern of media concentration in Australia is now found in many parts of the world including Britain, Europe, Canada and the United States ('Issue Guides: Media Concentration' 2003). The trend in media ownership is not only towards concentration within countries but also towards the creation of 'global media empires' that include newspapers, television stations, magazines, movie studios and publishing houses. For example, Murdoch also owns three of Britain's largest daily national newspapers and two of its largest circulation Sunday papers and controls extensive satellite broadcasting in dozens of countries. Murdoch's media empire also includes book publishing companies in Australia and the United States, Festival Records, and 20th Century Fox, as well as interests in computer software, offshore oil and gas and air transport (Abramsky 1995). According to journalist Sasha Abramsky, Murdoch 'has – and uses – the power to make' and break politicians and his papers 'have consistently opposed the peace movement, trade unions, progressive social programs . . . while supporting the death penalty, lower taxes at any cost and hawkish foreign policies' (1995: 16–17).

The mechanism of control generally exercised by media proprietors is through the appointment of editors, 'who become the proprietor's "voice" within the newsroom, ensuring that journalistic "independence" conforms to the preferred editorial line' (McNair 1994: 42). The power of the media is not just in its editorial line but also in covering some issues rather than others, some views but not others. It is this power that makes politicians so reluctant to cross the large media moguls and regulate the industry in the public interest. So while politicians would like to regulate against concentration of media ownership, they are not as tough as they would like to be on this score.

For liberal critics of the media, however, the business orientation of media owners and their relationship with other businesses is just as much a problem for media independence as the concentration of ownership in a few hands. Most media organisations are owned by multinational multi-billion dollar corporations that are involved in a number of businesses apart from the media, such as forestry, pulp and paper mills, defence, real estate, oil wells, agriculture, steel production, railways, water and power utilities (Kellner 1990: 82). Such conglomerates not only create potential conflicts of interest in reporting the news but ensure that the makers of the news take a corporate view.

The boards of these media companies typically include representatives of international banks, multinational oil companies, car manufacturers and other corporations. Noam Chomsky, who has documented a number of biases in the US media's treatment of foreign affairs, points out that media corporations 'are closely integrated with even larger conglomerates' and so it 'would hardly come as a surprise if the picture of the world they present were to reflect the perspectives and interests of the sellers, the buyers, and the product' (Chomsky 1989: 8).

The owners of the media influence the selection, shaping and framing of the news to attract advertisers – 'Proprietors determine the target audience and general editorial approach to that audience' (Windschuttle 1988: 264) – but also to ensure a favourable political climate for their media and other business concerns.

Commercial television and radio stations tend to get all their income from advertisers and newspapers are increasingly dependent on advertising. Tens of billions of dollars are spent every year just on television advertising and the media do their best to create a product that suits those advertisers. While audiences may consider the advertisements as an unwelcome interruption to their news and entertainment, in reality the news and entertainment are a way of attracting people to the medium so they will be exposed to the advertisements – a way of delivering audiences for advertisers (Parenti 1986: 62).

The influence of corporate advertisers on media content is both indirect, in that the media shape content to attract an audience that will suit its advertisers, and direct in that media outlets edit material that is likely to offend advertisers, especially with news stories (Franklin 1994: 43). Sometimes advertisers directly demand influence; one told *Time*, *Newsweek* and *US News & World Report* that it would give all its advertising to the magazine that gave the most favourable coverage to its industry (cited in Kilbourne 2000: 50). However, advertisers are not usually as blatantly upfront as that (see Jackson et al. 2003).

Corporations can also use sponsorship, a more indirect form of advertising, to influence the content of the media:

> Prospective shows are often discussed with major advertisers, who review script treatments and suggest changes when necessary. Adjustments are sometimes made to please sponsors . . . Corporate sponsors figure they are entitled to call the shots since they foot the bill – an assumption shared by network executives, who quickly learn to internalise the desires of their well-endowed patrons. (Lee & Solomon 1990: 60–1)

Large corporations that tend to sponsor newscasts and run green advertising campaigns were almost never examined for their environmental record (Letto 1995: 22).

KEY DEBATES

Journalistic objectivity

Journalists often claim that their own biases, and the pressures from advertisers and media owners, do not affect their work because of their professional norm of 'objectivity'. But the journalistic norm of objectivity is not the same as truth. It has three components. The first is 'depersonalisation', which means that journalists should not overtly express their own views, evaluations, or beliefs. The second is 'balance', which involves presenting the views of representatives of both sides of a controversy without favouring one side (Entman 1989: 30). And the third is 'accuracy', which requires journalists to quote people and relay 'facts' from sources accurately. And there are associated conventions:

> *authoritative* sources such as politicians must be quoted (in this way the journalist is seen to distance him- or herself from the views reported, by establishing that they are someone else's opinions); 'fact' must be separated from 'opinion', and 'hard news' from 'editorial comment'; and the presentation of information must be structured pyramidically, with the most important bits coming first, at the 'top' of the story. (McNair 1994: 47)

These conventions perpetuate the impression that reporters are simply conveying the 'facts' and not trying to influence how people interpret them. The ideal of objectivity gives journalists legitimacy as independent and credible sources of information.

The rhetoric of journalistic objectivity supplies a mask for the inevitable subjectivity that is involved in news reporting and is supposed to reassure audiences who might otherwise be wary of the power of the media. It also ensures a certain degree of autonomy to journalists and freedom from regulation to media corporations (Nelkin 1987: 94; Entman 1989: 32). However, news reporting involves judgements about what is a good story, who will be interviewed for it, what questions will be asked, which parts of those interviews will be printed or broadcast, what facts are relevant and how the story is written:

> value judgements infuse everything in the news media . . . Which of the infinite observations confronting the reporter will be ignored? Which of the facts noted will be included in the story? Which of the reported events will become the first paragraph? Which story will be prominently displayed on page 1 and which buried inside or discarded? . . . Mass media not only report the news – they also literally *make* the news. (Lee & Solomon 1990: 16)

Journalists are free to write what they like if they produce well-written stories 'free of any politically discordant tones', that is, if what they write fits the ideology of those above them in the hierarchy. A story that supports the status quo is generally considered to be neutral and its objectivity is not questioned, while one that challenges the status quo tends to be perceived as having a 'point of view' and therefore biased. Statements and assumptions that support the existing power structure are regarded as 'facts', while those that are critical of it tend to be rejected as 'opinions' (Parenti 1986: 35, 50). For example, one study of environmental stories found that 'while the media were willing to dispute dire environmental predictions, they were more accepting of dire economic projections – citing enormous anticipated job losses while rarely asking how the figures were derived, or if plant closings and layoffs were the only options' (Spencer 1992: 15).

Objectivity in journalism has nothing to do with seeking out the truth, except in so much as truth is a matter of accurately reporting what others have said. This contrasts with the concept of scientific objectivity where views are supposed to be verified with empirical evidence in a search for the truth (Nelkin 1987: 96). Ironically, journalistic objectivity discourages a search for evidence; the balancing of opinions often replaces journalistic investigation altogether. A survey of environmental reporting by media watchdog group Fairness and Accuracy in Reporting (FAIR) found that it tended to be 'limited to discussion of clashing opinions, rather than facts gathered by the reporters themselves' (Spencer 1992: 13).

News sources

The news is shaped by the choice of people journalists interview for research, quotes and on-air appearances. The conventions of objectivity, depersonalisation and balance tend to transform the news into a series of quotes and comments from a remarkably small number of sources. Most journalists tend to use, as sources, people from the mainstream establishment, whom they believe have more credibility with their audience. Highly placed government and corporate spokespeople are the safest and easiest sources in terms of giving stories legitimacy (Entman 1989: 18). When environmentalists are used as sources they tend to be leaders of the 'mainstream' environmental groups that are seen as more moderate (Spencer 1992: 17). Those without power, prestige and position have difficulty establishing their credibility as a source of news and tend to be marginalised (McNair 1994: 48).

Journalists who have access to highly placed government and corporate sources have to keep them on side by not reporting anything adverse about them or their organisations. Otherwise they risk losing them as sources of information. In return for this loyalty, their sources occasionally give them good stories, leaks and access to special interviews. Unofficial information, or leaks, gives the impression of investigative journalism, but is often a strategic manoeuvre on the part of those with position or power (Ricci 1993: 99). 'It is a bitter irony of source journalism . . . that the most esteemed journalists are precisely the most servile. For it is by making themselves useful to the powerful that they gain access to the "best" sources' (quoted in Lee & Solomon 1990: 18).

Balance means ensuring that statements by those challenging the establishment are balanced with statements by those they are criticising, though not necessarily the other way round (Parenti 1986: 52). For example, despite claims of anti-nuclear media bias by the nuclear industry, a FAIR study of news clippings collected by the Nuclear Regulatory Commission over a five-month period found that no news articles cited anti-nuclear views without also citing a pro-nuclear response, whereas 27 per cent of articles cited only pro-nuclear views (Grossman 1992).

Balance means getting opinions from both sides (where the journalist recognises two sides) but not necessarily covering the spectrum of opinion. More radical opinions are generally left out. Government environmental authorities can be used as an environmental source in one story and as an anti-environmentalist source in another. Nor are opposing opinions always treated equally in terms of space, positioning and framing (Parenti 1986: 218; Spencer 1992: 17). Balance does not guarantee neutrality even when sources are treated fairly, since the choice of balancing sources can be distorted. FAIR gives the example of a *Nightline* show where radio talk show host Rush Limbaugh argued that volcanoes are the major cause of ozone depletion. Limbaugh was 'balanced' with then Senator Al Gore, 'who argued that the answer to ecological problems was more "capitalism"' (Spencer 1992: 18).

'In practice objectivity means journalists have to interview legitimate elites on all major sides of a dispute' and this gives powerful people guaranteed access to the media no matter how flimsy their argument or how transparently self-interested. In their attempts to be balanced on a scientific story, journalists may use any opposing view 'no matter how little credence it may get from the larger scientific community' (Entman 1989: 37-8; Jim Naureckas, editor of *Extra!* quoted in Ruben 1994). But giving equal treatment to two sides of an argument can often give a misleading impression. Phil Shabecoff, former environment reporter for the *New York Times*, gives the example of views on climate change:

> the findings of the International Panel on Climate Change – a body of some 200 eminent scientists named by the World Meteorological Organization of the United Nations Environment Program – is generally considered to be the consensus position. But I have seen a number of stories where its conclusions are given equal or less weight than those of a single scientist who has done little or no significant peer-review research in the field, is rarely, if ever, cited on those issues in the scientific literature, and whose publication is funded by a fossil-fuel industry group with an obvious axe to grind . . . For a reporter, at this stage of the debate, to give equal or even more weight to that lonely scientist with suspect credentials is, in my view, taking sides in the debate. (1994: 42)

Paul Rauber gives another example of how equal treatment can give a misleading impression: 'Hundreds, maybe thousands of people gather to call for the factory to stop polluting or for the clearcutting to end. In one little corner, half a dozen loggers or millworkers hold a counter-demonstration on company time. That night on the evening news, both sides get equal coverage' (Rauber 1996: 20).

Journalists who accurately report what their sources say can effectively remove responsibility for their stories onto their sources. The ideal of objectivity therefore encourages uncritical reporting of official statements and those of authority figures. In this way the biases of individual journalists are avoided but institutional biases are reinforced (Ryan 1991: 10, 176). 'Professional codes ensure that what is considered important is that which is said and done by important people. And important people are people in power. Television news thus privileges holders of power' (Kellner 1990: 113–14).

Front groups and think-tanks

Powerful corporations are not only represented in the media by corporate spokespeople but they also seek to multiply their voice by funding others to speak for them as well. A major focus of the new corporate activism, which has been a response to the perceived liberal bias of journalists, has been to ensure that corporate-funded people are the ones that the media turn to for comment, be they scientists, think-tank 'experts' or front group spokespeople.

The use of front groups enables corporations to take part in public debates in the media behind a cover of community concern. When a corporation wants to oppose environmental regulations or support an environmentally damaging development it may do so openly and in its own name. But it is far more effective to have a group of citizens or a group of experts – preferably a coalition of such groups – which can publicly promote the outcomes desired by the corporation while claiming to represent the public interest. When such groups do not already exist, the modern corporation can pay a public relations firm to create them.

Merrill Rose, executive vice-president of the public relations firm Porter/Novelli, advises companies:

> Put your words in someone else's mouth . . . There will be times when the position you advocate, no matter how well framed and supported, will not be accepted by the public simply because you are who you are. Any institution with a vested commercial interest in the outcome of an issue has a natural credibility barrier to overcome with the public, and often, with the media. (Rose 1991)

Corporate front groups often portray themselves as environmentalists. In this way corporate interests appear to have environmental support. The names of these groups are chosen because they sound as if they are grassroots community and environmental groups. The Forest Protection Society in Australia, for example, was established in 1987 with the support of the Forest Industry Campaign Association (Rowell 1996: 240). It shared the same postal address as the National Association of Forest Industries and its fact sheets promote logging in rainforests as 'one of the best ways to ensure that the rain-forests are not destroyed' (Burton 1994: 17–18). (In 2000 it came out of the closet and renamed itself Timber Communities Australia [http://www.tca.org.au/TCAIndex.htm].)

Corporate front groups may also portray themselves as independent scientific groups whose aim is to cast doubt on the severity of the problems associated with environmental deterioration and create confusion by magnifying uncertainties and showing that some scientists dispute the claims of the scientific community. For example groups funded by the fossil-fuel industry emphasise the uncertainty associated with global warming predictions (Beder 2002b: ch. 14).

Another strategy used by corporate front groups is to recognise environmental problems caused by corporations but to promote superficial solutions that prevent and pre-empt the sorts of changes really necessary to solve the problem. Sometimes they shift the blame from corporations to the individual citizen. For example, the Keep America Beautiful Campaign focuses on anti-litter campaigns but ignores the potential of recycling legislation and changes to packaging. It seeks to attribute litter and waste disposal problems to individuals acting irresponsibly and admits no corporate responsibility (Beder 2002b: 30).

The media often use these front groups as sources of information and quote their spokespeople without realising their corporate origins or acknowledging in their news reporting the corporate connections of the groups. The same is true for think-tanks, which are overwhelmingly funded by corporations and wealthy corporate-aligned foundations. Various studies by FAIR have found that conservative and centrist think-tank experts are used as news sources many times more often than experts from progressive or left-leaning think-tanks (Dolny 2000). These think-tanks are cited without any indication of their ideological basis or funding sources and their personnel are treated as independent experts (Solomon 1996: 10).

The increasing trend for corporations to use front groups and friendly scientists as their mouthpieces has distorted media reporting on environmental issues since the media often do not differentiate between corporate front groups and genuine citizen groups, and industry-funded scientists are often treated as independent scientists. Because of the myth of scientific objectivity journalists tend to have an uncritical trust in scientists (Nelkin 1987: 105) and few 'question the motivation of the scientists whose research is quoted, rarely attributing a study's funding source or institution's political slant' (Ruben 1994: 11). Nor do the mainstream media generally cover the phenomenon of front groups and think-tanks and artificially generated grassroots campaigns, which would serve to undermine their operation by exposing the deceit on which they depend.

Public Relations

Much of the news people read or watch on television is manufactured by PR firms and specialists, rather than discovered by journalists. Media and press releases include news, feature stories, bulletins, media advisories and announcements, all of which flood media offices. Their purpose is to develop and maintain public goodwill for the organisation sending them as well as favourable government policies. Most journalists rely on these sources to supply the 'raw material of their craft, regular, reliable and useable information' (Walters & Walters 1992: 33). This flow of 'free' information saves the journalist time and effort finding stories to write about. Yet it is very difficult for the public to be able to distinguish real news from news generated by public relations.

Often news stories are copied straight from news releases; at other times they are rephrased and sometimes they are augmented with additional material. This practice does not vary much between large and small papers as larger papers need more stories and smaller papers have fewer staff to write their stories. According to various studies, press releases are the basis for 40–50 per cent of the news content of US newspapers (Blyskal & Blyskal 1985: 28; Lee & Solomon 1990: 66; Walters & Walters 1992: 33; Carlisle 1993: 22).

The reliance of journalists on sources such as PR personnel and government officials is referred to as 'source journalism'. By providing the news feedstock, they cause reporters to react rather than initiate. Journalists who are fed news stories are less likely to go looking for their own stories, which could bring negative publicity. In this way source journalism displaces investigative reporting.

By being the primary source of a journalist's information on a particular story, PR people can influence the way the story is told and who tells it. Jeff and Marie Blyskal in their book *PR: How the Public Relations Industry Writes the News* explain:

> Good PR is rather like the placement of a fish-eye lens in front of the reporter. The facts the PR man wants the reporter to see front and center through the lens appear bigger than normal. Other facts, perhaps opposing ones, are pushed to the side by the PR fish-eye lens and appear crowded together, confused, obscured. The reporter's entire field of vision is distorted by the PR lens. (Blyskal & Blyskal 1985: 69)

Public relations is a multi-billion dollar industry. In 2000, the top twenty-five public relations companies received over $3600 million in revenues and in the US alone employed over 200 000 people (Holmes Report 2001). One of the fastest growing areas of public relations is environmental public relations, or 'greenwash' as environmentalists call it. Between 1990 and 1995 the amount that US firms were spending each year on public relations advice on how to green their own image and deal with environmental opposition doubled to about $1 billion per year (Bleifuss 1995: 4–9; Stauber & Rampton 1995: 173). Today most of the top PR firms include environmental PR as one of their specialities.

One of the ways PR experts enhance the image of their clients and show that they care is by emphasising and publicising their positive actions, no matter how trivial, and downplaying any negative aspects, no matter how significant. According to Robert Gray, former chairman of PR giant Hill & Knowlton, 'our job is not to make white black or to cover the truth, but to tell the positive side regardless of who the client is' (quoted in Roschwalb 1994: 270). Sometimes this involves putting a positive spin or interpretation on the available information:

> Did this year's fines levied by the Environmental Protection Agency (or the state equivalent) drop to 'only' $5 million? Then celebrate the company's 'continued positive trend in compliance.' Was there no improvement from last year's release of toxic chemicals? Then report on the 'levelling off of emissions.' (Makower 1996)

For example, in March 1999 BP launched its 'Plug in the Sun' program based on its investment in solar energy and the installation of solar panels on petrol stations around the world. In its advertisements it said, 'We can fill you up by sunshine' although it was still petrol people were putting in their cars. In 2000 it rebranded itself as 'bp, beyond petroleum'. But bp remains committed to ever-increasing production and usage of oil and gas and it spent more on its rebranding than it did on solar energy (quoted in Beder 2002a).

News style

Environmental problems are poorly reported in the media because of the need to provide entertainment rather than political awareness, to attract audiences for advertisers, even in news and current affairs programs. This occasionally affects a specific item of news but more generally affects the sorts of stories that are covered and the way they are covered. News editors are reluctant to deal with controversial political and social issues that might alienate potential consumers. As a result news has become bland and neutral and ignores issues that concern large portions of the population who are not considered to have or exercise much buying power (Bagdikian 1983: 180–1, 201–2).

Yet bland news can be boring, so the lack of controversy and social significance is made up for by making the news entertaining and interesting. Intellectual and political interest is replaced by 'human interest', conflict, novelty, emotion and drama, or as one feature writer put it, 'currency, celebrity, proximity, impact and oddity' – the elements of newsworthiness (Ryan 1991: 31).

Entertainment merged with current affairs produces 'infotainment' which, as Philip Gold notes in the conservative magazine *Insight on the News*, blends 'trivial amusement with the address of serious issues', reduces 'serious reportage into fragmented coverage of the latest "shocking developments"' and squeezes out 'more serious discourse' (Gold 1994: 37–8). Television news producers prefer very short stories with good visuals and action stories that add excitement to the news. They are very good at providing drama and emotion but poor at giving in-depth information on complex issues. News stories are presented very quickly, in rapid succession and with little explanation. As a result, people who rely on television to get their news tend to be 'the least-informed members of the public' (Levy 1992: 70).

The need to entertain turns social processes and events into stories. Stories that take longer than a day to unfold are told as a series of climaxes (Windschuttle 1988: 268). Says one editor: 'Acid rain, hazardous waste . . . they're the kind of big bureaucratic stories that make people's eyes glaze over. There's no clear solution, no clear impact. They're not sexy' (quoted in Ryan 1991: 31). The news 'is characteristically about events rather than processes, and effects rather than causes' (McNair 1994: 46). As a result environmental reporting tends to concentrate on events such as the Earth Summit or various Earth Days, accidents, disasters such as oil tanker spills, and official announcements (Spencer 1991: 13) and avoids background information on context and structural causes.

News stories are told as 'self-contained, isolated happenings' (Gamson et al. 1992: 387). Reporting of environmental problems tends to be superficial, narrowing the focus to specific events in isolation rather than looking at systemic problems that caused them such as the international monetary system or the unregulated power of corporations, and concentrating on the costs of environmental measures (Lee & Solomon 1990: 202, 222). Environmental problems become a series of events that emphasise individual action rather than social forces and issues.

Each story competes for priority and an emphasis on 'breaking news' does not encourage any coverage of long-term issues. This means journalists have to work to very tight deadlines and don't have the time to investigate properly and consult a wide range of sources.

The journalistic tendency to balance stories with two opposing views leads to a tendency to 'build stories around a confrontation between protagonists and antagonists' (Ricci 1993: 95). Issues such as garbage and sewage sludge only get coverage, despite their importance, when there is a fight over the siting of a landfill or incinerator and then the coverage is on the 'anger and anguish of affected citizens, or the conflicting claims of corporate spokesmen, government regulators and environmental activists' rather than the issues and technical background to them (Gersh 1992: 16).

The environmental movement relies extensively on the mass media to get its message across to the general public, but doing so has its costs. The media tend to present images and style, not meaning and content. Protest actions and events are described as theatre spectacles rather than as 'part of a democratic struggle over vital issues' (Parenti

1986: 99). In his analysis of how the media treated the New Left student movement of the 1960s and 1970s, Todd Gitlin (1980: 3) observed:

> In the late twentieth century, political movements feel called upon to rely on large-scale communications in order to *matter*, to say who they are and what they intend to publics they want to sway; but in the process they become 'newsworthy' only by submitting to the implicit rules of newsmaking, by conforming to journalistic notions (themselves embedded in history) of what a 'story' is, what an 'event' is, what a 'protest' is. The processed image then tends to become 'the movement' for wider publics.

In many news stories about local controversies the intelligence and research of local residents is downplayed and they are presented as passionate, self-interested and inexpert. This tends to discourage wider support for their cause from the viewing public and to disempower other citizens by depriving them of attractive models of political activism.

Current affairs programs do expose corporate misdeeds, accidents and environmental and health problems resulting from unsafe products and production processes but in a way that does not call into question 'fundamental political or economic structures and institutions' (Kellner 1990: 107–8).

> By treating business wrongdoings as isolated deviations from the socially beneficial system of 'responsible capitalism', the media overlook the systemic features that produce such abuses and the regularity with which they occur. Business 'abuse' is presented in the national press as an occasional aberration, rather than as a predictable and common outcome of corporate power and the business system. (Parenti 1986: 110–11)

Environmental disasters are not followed up and environmental revelations that are uncovered by journalists are 'seldom incorporated into the body of knowledge and perspective' that environmental journalists draw on in their work (Spencer 1992: 13).

The environmental movement is often characterised in the media as 'just another special-interest group' looking after its own 'economic and institutional well-being' rather than a 'broad-based social movement' (Shabecoff 1994: 43). The more radical environmental groups are sometimes treated as fringe loonies.

FUTURE DIRECTIONS

The shaping of the news is a well-understood area despite the existence of differing views on the significance and impact of news filters such as journalists, media owners, advertisers and news sources. The implications of these filters for the reporting of environmental news stories can readily be deduced. But the wider implications for environmentalists seeking to change perceptions of environmental problems and encourage action to be taken to solve them is an area that requires further research and discussion.

Discussion Questions

1. Why does concentration of media ownership matter? Does it affect environmental news coverage?
2. Do advertisers influence news content? In what way might this affect how environmental issues are covered in the news?
3. How objective can news reporters be? Is a concern for the environment objective?
4. How do journalists find their news sources? How would an environmental group go about becoming a news source?
5. How can you differentiate between a front group, and a genuine community-based environmental group?
6. What is wrong with source journalism? Why do environmentalists prefer investigative reporting?
7. Why does it matter if television stations try to make the news entertaining? How does it influence the reporting of environmental issues?
8. What are the filters that an issue or event has to pass through to become a news story? Which filters are most significant for environmental reporting?

Glossary of Terms

Front groups: a group that purports to be independent or broadly community-based but is in fact funded or sponsored to represent a special interest.

Greenwash: an environmental adaptation of 'whitewash' aimed at pretending that an organisation or person cares about the environment.

News source: people whom journalists interview as part of their research or for quotes and on-air appearances.

Source journalism: journalism that relies on information provided by sources and from news releases and public relations personnel.

Think-tanks: organisations that are oriented towards propagating particular research and ideas that are in the interests of their funders, which are usually corporations.

References

Abramsky, S. 1995, 'Citizen Murdoch: the shape of things to come?' *Extra!* Nov.–Dec.: 16–17.
Alterman, E. 2003, *What Liberal Media? The truth about bias and the news*, New York: Basic Books.

Bagdikian, B.H. 1983, *The Media Monopoly*, Boston MA: Beacon Press.

Beder, S. 2002a, 'Bp: Beyond Petroleum?' In Eveline Lubbers (ed.) *Battling Big Business: Countering greenwash, infiltration and other forms of corporate bullying*, Devon, UK: Green Books, pp. 26–32.

Beder, S. 2002b, *Global Spin: The Corporate assault on environmentalism*, rev. edn, Totnes, Devon, UK: Green Books.

Bleifuss, J. 1995, 'Covering the earth with "Green PR"', *PR Watch* 2(1): 1–7.

Blyskal, J., and M. Blyskal 1985, *PR: How the public relations industry writes the news*, New York: William Morrow & Co.

Burton, B. 1994, 'Nice names – pity about the policies – industry front groups', *Chain Reaction* (70): 16–19.

Carlisle, J. 1993, 'Public Relationships: Hill and Knowlton, Robert Gray, and the CIA', *Covert Action* (44): 19–25.

Chomsky, N. 1989, *Necessary Illusions: Thought control in democratic societies*, London: Pluto Press.

Dolny, M. 2000, 'Think tanks: the rich get richer', *Extra!* May–June.

Entman, R.M. 1989, *Democracy without Citizens: Media and the decay of American politics*, New York: Oxford University Press.

Franklin, B. 1994, *Packaging Politics: Political communications in Britain's media democracy*, London: Edward Arnold.

Gamson, W.A., D. Croteau, W. Hoynes and T. Sasson 1992, 'Media Images and the social construction of reality', *Annual Review of Sociology* 18: 373–93.

Gersh, D. 1992, 'Covering solid waste issues', *Editor & Publisher* 125 (29 August): 15–16.

Gitlin, T. 1980, *The Whole World Is Watching: Mass media in the making and unmaking of the New Left*, Berkeley CA: University of California Press.

Gold, P. 1994, 'Just Say no to infotainment', *Insight on the News* 10(28): 37–8.

Goldberg, B. 2003a, *Arrogance: Rescuing America from the media elite*, New York: Warner Books.

Goldberg, B. 2003b, *Bias: A CBS Insider exposes how the media distort the news*, New York: Perennial.

Grossman, K. 1992, 'Survey Says: Newspapers Boost Nukes', *Extra!* March: 14.

Herman, E.S., and N. Chomsky 2002, *Manufacturing Consent: The political economy of the mass media*, New York: Pantheon Books.

Holmes Report 2001, Council of PR Firms Rankings. http://www.holmesreport.com/agencies/rankings.cfm (accessed 10 March 2004).

'Issue Guides: Media Concentration' 2003, Mediachannel.org. http://www.mediachannel.org/ownership/front.shtml (accessed 1 December).

Jackson, J., P. Hart and R. Coen 2003, 'Fear and favor 2002 – the third annual report', *Extra!* March–April. http://www.fair.org/extra/0303/fear-favor-2003.html.

Kellner, D. 1990, *Television and the Crisis of Democracy*, Boulder CO: Westview.

Kilbourne, J. 2000, *Can't Buy My Love: How advertising changes the way we think and feel*, New York: Touchstone.

Kuypers, J.A. 2002, *Press Bias and Politics: How the media frame controversial issues*, Westport CT: Praeger Publishers.

Lawson, A. 2003, 'Murdoch faces Australian setback', *Guardian* 25 June.

Lee, M.A., and N. Solomon 1990, *Unreliable Sources: A guide to detecting bias in news media*, New York: Carol Publishing Group.

Letto, J. 1995, 'TV lets corporations pull green wool over viewers' eyes', *Extra!* July–August: 21–4.

Levy, M.R. 1992, 'Learning from television news'. In P.S. Cook, D. Gomery and L.W. Lichty (eds) *The Future of News: Television-newspapers-wire services-newsmagazines*, Washington DC: Woodrow Wilson Center Press, pp. 69–71.

Makower, J. 1996, 'Just the Facts', *E: the Environmental Magazine* 7(2): 48, 50.

McNair, B. 1994, *News and Journalism in the UK*, London/New York: Routledge.

Nelkin, D. 1987, *Selling Science: How the press covers science and technology*, New York: W.H. Freeman & Co.

Parenti, M. 1986, *Inventing Reality: The politics of the mass media*, New York: St Martin's Press.

Parry, R. 1995, 'The Rise of the Right-Wing Media Machine', *Extra!* March/April: 6–10.

Phillips, P., N. Chomsky and T. Tomorrow 2001, *Censored 2001: 25 years of censored news and the top censored stories of the year*, New York: Seven Stories Press.

Rauber, P. 1996, 'The Uncertainty Principle', *Sierra* 81 (Sept.–Oct.): 20–2.

Ricci, D. 1993, *The Transformation of American Politics: The new Washington and the rise of think tanks*, New Haven CT: Yale University Press.

Roschwalb, S.A. 1994, 'The Hill & Knowlton cases: a brief on the controversy', *Public Relations Review* 20(3): 267–76.

Rose, M. 1991, 'Activism in the 90s: changing roles for public relations', *Public Relations Quarterly* 36(3): 28–32.

Rowell, A. 1996, *Green Backlash: Global subversion of the environment movement*, London/New York: Routledge.

Ruben, B. 1994, 'Back Talk', *Environmental Action Magazine* 25(4): 11–16.

Ryan, C. 1991, *Prime Time Activism: Media strategies for grassroots organizing*, Boston MA: South End Press.

Saloma, J.S. 1984, *Ominous Politics: The new conservative labyrinth*, New York: Hill & Wang.

Shabecoff, P. 1994, 'Mudslinging on the earth-beat', *Amicus Journal* 15(4): 42–3.

Solomon, N. 1996, 'The media's favorite think tank', *Extra!* July/August: 9–12.

Spencer, M. 1991, 'Cold war environmentalism: reporting on Eastern European pollution', *Extra!* January/February: 6–7.

Spencer, M. 1992, 'U.S. environmental reporting: the big fizzle', *Extra!* April–May: 12–22.

Stauber, J.C., and S. Rampton 1995, '"Democracy" for hire: public relations and environmental movements', *Ecologist* 25(5): 173–80.

Walters, L.M., and T.N. Walters 1992, 'Environment of confidence: daily newspaper use of press releases', *Public Relations Review* 18(1): 31–46.

West, D.M. 2001, *The Rise and Fall of the Media Establishment*, Boston MA: Bedford/St Martin's Press.

Windschuttle, K. 1988, *The Media: A New Analysis of the Press, Television, Radio and Advertising in Australia*, 2nd edn, Melbourne: Penguin.

CHAPTER THIRTEEN

AGRICULTURAL PRODUCTION AND THE ECOLOGICAL QUESTION

Geoffrey Lawrence, Lynda Cheshire and Carol Ackroyd Richards

Environmental degradation is a worldwide phenomenon. It is manifested in the clearing of forests, polluted waterways, soil erosion, the loss of biodiversity, the presence of chemicals in the ecosystem and a host of other concerns. Modern agricultural practices have been implicated in much of this degradation. This chapter explores the connections between the form of agricultural production undertaken in advanced nations – so-called 'productivist' or 'high-tech' farming – and environmental degradation. It is argued that the entrenchment of productivist agriculture has placed considerable, and continuing, pressures on the environment and, second, that while no new options for a more sustainable agriculture and new policies are being proposed to tackle the existing problem, the underlying basis of productivist agriculture remains largely unchallenged. The prediction is that environmental degradation will continue unabated until more dramatic (and possibly less palatable) measures are taken to alter the behaviour of producers and the trajectory of farming and grazing industries throughout the world.

The 'ecological question' we pose is straightforward: is it possible for modern agriculture – one currently wedded to a regime of petrochemicals, pesticides, weedicides, insecticides, artificial fertilisers, crop monocultures and intensive animal production – to become sustainable?

BACKGROUND TO THE ISSUES

Environmental issues have been recognised throughout recent history. Soil fertility depletion was a major concern, for example, in Europe and North America during the forty years of the mid-1800s, prompting social thinkers like Karl Marx to theorise the role of capitalism in environmental decline (Foster & Magdoff 1998). Observing that soil nutrient cycles were being compromised by the profit-making desires of farmers, and that hired farm labour was – like other labour – exploited as surplus value to be appropriated by employer-farmers, Marx wrote that progress in capitalist agriculture was characterised by exploitation both of workers and the environment. In considering how this would be overcome, Marx argued that sustainable production would only be possible with a change in the ownership and control of production:

> From the standpoint of a higher socio-economic formation, the private property of particular individuals in the earth will appear . . . absurd . . . Even an entire society, a nation, or all simultaneously existing societies taken together, are not owners of the earth, they are simply its possessors, its beneficiaries, and have to bequeath it in an improved state to succeeding generations as [good heads of the household]. (cited in Foster & Magdoff 1998: 38)

Here Marx is highlighting what some have called a 'stewardship ethos': a recognition by rural producers that they have an obligation to future generations to preserve the land and hand it on in a non-degraded state. Yet this is not occurring and sociologists, among others, have raised questions about why this might be. Many, inspired by Marx's writings (Davis 1980; Bonanno et al. 1994; Heffernan 1998), have sought to explain the problem in terms of the structural position of family farmers in a world of corporate capital. They argue that farmers own small amounts of capital and occupy what some have viewed as a 'contradictory' class position, which places them in a vulnerable situation (Mooney 1988). Neither capitalist (employing large amounts of finance and labour) nor labourer (working solely for a wage), farmers use their own, and family, labour to work with relatively small amounts of capital, and are often unable to meet the costs of production from the sale of agricultural produce. In many advanced nations, the state subsidises agriculture to ensure returns are at a reasonable level (Lawrence 1987). In others, where neo-liberal policies favour exposure of farmers to free-market conditions, they must employ a number of strategies to 'survive' and grow. Being market-oriented they tend to *specialise* in production, putting their know-how and capital into one or several commodities. They *intensify* their operations, using external capital and other inputs to increase productivity and efficiency. And finally, they also seek to expand the size of their operations in order to take advantage of returns to scale. The result is that fewer farmers produce ever-increasing amounts of food and fibre: something referred to as economic *concentration* (Heffernan 1998).

Specialisation, intensification and concentration are three of the main features of productivist agriculture (Ilbery & Bowler 1998; Argent 2002). Productivist agriculture is praised by many in the West, and is viewed as the only system of farming likely to solve the world 'food crisis' (Avery 1995). Food and fibre are produced in a highly efficient manner, using the latest technologies, and delivered to consumers throughout the world year-round. In the Third World, the combination of high-tech inputs, irrigation, machinery and the latest hybrid varieties lead to an explosion in the production of agricultural commodities. Much of this is exported to the West, providing, so the story goes, much-needed foreign income to developing nations.

Despite the supposed advantages of productivist agriculture, there are now well-acknowledged problems that have accompanied its expansion. Five main concerns relate to: the control of the food system by transnational capital; the removal of people from the land as they drift to the cities in search of employment (particularly in developing nations); the demise of rural communities; the failure of the present system of agricultural production and distribution to feed the poor and malnourished peoples of the world; and widespread environmental degradation (see Magdoff et al. 1998). It is this final issue that will be dealt with in this chapter.

How, then, might productivist agriculture compromise the environment? First, the specialisation of producers into crop production does much to prevent recycling as producers without access to natural fertilisers are forced to purchase synthetic fertilisers and other chemicals to use over large areas. At the individual farm level there may not appear to be a problem, yet the cumulative run-off from the fertilisers and pesticide sprays enters waterways and pollutes river systems. Second, while pest infestations might be temporarily prevented by chemical applications, they quickly develop resistance, with farmers responding by adopting more potent chemicals in the next production period. Third, although synthetic fertilisers are employed to increase output, they do not improve soil structure. The result is that farmers are forced to increase levels of external, synthetic inputs over time, placing them, in this sense, on a 'treadmill of production' (Schnaiberg 1980). Fourth, the production of synthetic fertilisers (created from non-renewable resources) contributes to global warming (Altieri 1998). Fifth, land clearing for agriculture and grazing leads to increased loss of topsoils and, as water tables rise, to the salinisation of millions of hectares of those remaining (National Farmers' Federation and Australian Conservation Foundation 2000). Farmers are also overcropping and overgrazing in order to remain in production, hoping that better seasons (and prices) will come (see Richards et al. 2003). Such environmentally damaging behaviour is a worldwide phenomenon in global market-dependent nations, and/or nations with poor regulatory regimes (see M. O'Connor 1994a; Martinez-Alier 1999).

While attempts to arrest environmental damage caused by productivist agriculture might appear to lie in the application of better science, this chapter argues that technological fixes to environmental degradation often form part of the problem and not the solution. In this case, it is necessary to ask how else we might approach the ecological question, and to examine the contribution that sociology can make to a better understanding of the relationship between modern agriculture and environmental damage. For the past two decades or more, sociologists and other social scientists have convincingly argued that many environmental problems have social dimensions that cannot be addressed, or even understood, by soil scientists, hydrologists and engineers. This is because agriculture and resource management are inherently social activities, based on particular kinds of social relationships that take place between people and the environment. Moreover, these relationships occur within the context of much broader structural imperatives and power dynamics that shape and influence the kind of relationships farmers and other producers are able to establish with the environment, and with other actors within the agri-food regime.

Rather than seeking individualistic, psychological explanations for on-farm environmental damage, sociologists have attempted to locate farmers' actions within the context of advanced capitalism by exploring the conditions that induce them to engage in unsustainable farming practices, even when they embrace a stewardship ethos. They have also sought to examine the ways in which truth and knowledge are socially constructed by various actors in the agri-food arena, each of whom has competing interests and differential access to power and resources. As well, sociologists have refused to take commonsense or taken-for-granted concepts at face value, and have argued for the need to 'unpack' what is meant by such terms as 'the environment' and 'sustainability'.

The very means by which sociologists conceive and explore the question of environmental sustainability has been a matter of debate within the discipline itself. Environmental sociologists, for example, have tended to adopt a more materialist conception of the social and natural worlds, thereby viewing the environment in a 'singular, undifferentiated, sense as an overarching "thing" or "whole" which is tending to be progressively degraded and polluted' (Buttel 2001: 24). Those adopting a social constructivist perspective, on the other hand, acknowledge that the environment is a symbolic or cultural, as well as material, artefact whose meaning is constructed through interaction and discourse. From this perspective attitudes, values and dispositions are recognised as crucial in shaping how people perceive the natural world and the nature and extent of its destruction.

KEY DEBATES

Given the diversity that characterises the sociological discipline, there is a broad range of debates in which sociologists are engaging about the nature, extent and possible solutions to the problems of environmental degradation that arise from agricultural production. Since not all can be addressed here, our discussion is limited to four key areas of concern:

1. How can we explain land degradation as an outcome of productivist agriculture? Are farmers to be held responsible for failing to manage their land in an appropriate manner or can we look to broader, structural conditions that compromise farmers' ability to engage in alternative, and perhaps more sustainable, farming practices?

2. If, as we argue, there is a strong connection between environmental degradation and productivist agriculture, is it possible for sustainability to occur within the parameters of productivism or is a more deep-seated change to the trajectory of capitalist advancement required?

3. What policy settings are expected to bring about change? What insights can sociologists bring to debates about regulatory, compensatory and self-help solutions to environmental degradation at the farm level?

4. Are low-input and organic production systems a viable option in the future for agriculture?

Explaining land degradation

Despite the problems that agricultural production has caused to the environment, it has been well documented that most landholders subscribe to a stewardship ethic (see Vanclay & Lawrence 1995). There is a strong sustainability assertion in this, and it has been reasoned that if all producers held firmly to a stewardship ethos, then we would be well on the way to a sustainable future – at least in relation to rural production. If this is the case, why has rural production caused such extensive environmental degradation? In relation to the farming population, holding these ideals appears to be a poor predictor of behaviour that might, in a practical manner, lead to better on-farm management (Vanclay & Lawrence 1995; Cary et al. 2002).

Given that many primary producers purport to be guardians of the land, are they to blame for the impacts of their land management practices on the environment? Until sociology took up the challenge of natural resource management, a considerable body of work examined 'the farmer' as an individual who chose to either adopt or reject new practices. Thus researchers described farmers as 'adopters', 'non-adopters' or 'laggards' according to their willingness to embrace new methods of production (Rogers 1962). While such analyses highlighted the social aspects of environmental management, they failed to acknowledge landholders as actors within complex social, economic and political structures. In other words, 'failing to adopt' a new method of farming or land management may not be contingent on whether the farmer has the 'right attitude' (see Lockie 2001), but rather on a complex interplay of local and global issues such as local community values, climate, farm economic viability, commodity prices, the agri-food requirements of transnational corporations, trade agreements and so on.

There are, in fact, a number of structural constraints that, despite good intentions and an ethic of stewardship, lock primary producers into a system of management that is not sustainable in the long term. Indeed, landholders remain at the mercy of free-trade agreements, low commodity prices and competition with nations who subsidise agricultural production (see Gray & Lawrence 2001). Those striving to remain economically viable often have to do so at the expense of the environment, for example by increasing chemical inputs and clearing trees to maximise the productive capacity of the property.

While it appears rational to blame farmers for exploiting the environment, sociologists have questioned some of the assumptions that underpin debates about farmers' environmental responsibilities (Lockie 2001). As many landholders point out, moves to more sustainable farming methods are costly, yet farmers are expected to bear these costs alone – in spite of the greater social good that results from these practices (Gray & Lawrence 2001). Similarly, the procurement of cheap agri-food products by fast food chains with considerable global market presence and buying power pitches primary producers against one another in the competition for contracts (Burch & Rickson 2001). The low cost of these commodities enjoyed by the fast food industry is ultimately subsidised by the environment.

Sustaining productivist agriculture

There are few who would disagree that a shift to a more sustainable form of agriculture is both desirable and necessary. Precisely what is meant by sustainability, and how we might achieve this goal, however, are matters of considerable disagreement that are influenced, as much as anything, by the goals and interests of relevant stakeholders, and by the power of those stakeholders to articulate their own truth-claims at the expense of others. As sociologists, it is important to take a critical look at competing discourses of sustainability to examine the ways in which they lend support to, or fundamentally challenge, the existing structures and practices of capitalism. Indeed, one important question that arises from such analysis is whether agriculture can become more sustainable without any radical overhaul of the capitalist system, or whether there are in fact inherent contradictions to capitalism that render a move towards sustainable farming practices unlikely, if not impossible.

James O'Connor, for example, argues that environmental degradation is inevitable under capitalism due to the existence of two contradictions inherent in the capitalist imperative. The first of these contradictions – what O'Connor calls a 'demand crisis' (1994: 160) – arises where individual producers seek to increase efficiency and profits by shedding labour, intensifying production practices, or employing new (usually expensive) technologies. When all, or most, producers do this, the volume of output increases, resulting in some circumstances in 'overproduction'. Overproduction drives down the prices that are received by farmers, subjecting them to a cost–price squeeze as the cost of farm inputs continues to rise.

It is under such conditions that farmers are usually forced to resort to measures that, while increasing production in the short term, have long-term environmental consequences. Here, we refer to the second contradiction of capitalism: an emerging 'cost crisis' (J. O'Connor 1994: 162), which arises when producers exploit the land until its productivity is severely reduced. When this occurs, it becomes even more costly to maintain, let alone increase, production as soils and waterways become poisoned, pests become resilient to the chemicals being used on the farm, and grazing pastures become eroded and bare. For farmers, the outcome of the cost crisis is twofold. Not only does it place them in a position that forces them to further intensify their often harmful farming practices, but it also means that future generations will inevitably inherit a degraded natural resource. In effect, farmers are required to make a choice about the trade-off between long-term ecological sustainability and short-term profit (Stoneham et al. 2003: 197). Many do attempt to adopt restorative practices to improve on-farm land and water quality (Lockie 2001: 238) yet, in the absence of appropriate regulation from the state, the capacity of producers to adopt more sustainable farming practices is severely curtailed.

Examining policy solutions

If the state continues to be a major supporter of productivist agriculture, and if environmental degradation poses a challenge to the continuation of the capitalist process, then it is clearly in the interest of political authorities to ensure something is done to arrest environmental decline (Stoneham et al. 2003: 197). In light of growing public concerns about the ecological impacts of agriculture, social scientists have suggested that governments are likely to face a legitimacy crisis unless they are seen to be taking remedial action (J. O'Connor 1994; Lockie 2000). These two contradictory imperatives – maintaining accumulation *and* legitimacy – have prompted political authorities to adopt a broad range of policy measures designed to encourage a more sustainable form of agriculture. The contribution of social scientists to policy debates about the form this sustainable agriculture might take has occurred in two forms. On the one hand sociology has played a key role in the formulation and implementation of policy responses, initially being 'tacked on' to the natural sciences. More recently, however, the insights offered by social scientists have become much more critical and theoretically inspired. As a result, many of the current policy solutions advanced by contemporary political authorities have been rendered problematic and shown to prioritise market imperatives for increased productivity at the expense of ecological sustainability.

In essence, the broad range of policies and programs that have been advanced by political authorities and analysed by sociologists can be grouped into three main strategies. The first of these incorporates the now discredited 'transfer of technology' approach so strongly favoured during the 1970s and 1980s. While social scientists have long since pointed out that attempts to address environmental decline by training farmers to adopt better farming practices are overly 'top-down' (Wright 1990: 44) and 'technocratic' (Vanclay 1997: 12), there is still a widely held view that the causes of, and hence the solutions to, environmental degradation arise overwhelmingly from the attitudes and practices of producers themselves. What were once seen as collective problems for the state and/or society are now regarded as individual problems to be addressed through strategies of self-help at the farm or community level (Herbert-Cheshire 2000; Higgins 2002). Internationally, Australia's National Landcare Program, which rose to prominence on a noble platform of community ownership and empowerment (Martin 1997: 45), is one of the most significant manifestations of this individualisation process. However, Landcare has also been subject to extensive criticism by sociologists and accused of being overly reliant on the skills and commitment of local volunteers (Lockie & Vanclay 1997). The effect, they argue, is that Landcare and other self-help approaches are not simply 'band aid' remedies to a deep-seated malaise, but work in the interests of capitalist expansion by shifting the blame for environmental degradation firmly onto the shoulders of local people while leaving the structural imperatives of productivist agriculture unchallenged (Lockie 2000; Herbert-Cheshire 2001).

The second range of policy instruments currently being used represents an attempt by state agencies to respond to criticisms that the urgently needed shift to sustainable farming is unlikely to occur through voluntary action. What is required, they argue, are regulatory mechanisms that enforce a change of practice, either by penalising or prosecuting non-compliance, or by providing farmers with financial inducements to adopt new methods. Examples of these kinds of frameworks currently being adopted in Australia and elsewhere include the imposition of levies on land used for particular crops such as annual or mono-crops, tax rebates for farmers investing in recommended forms of production, and penalties for those who overuse water or clear vegetation (Industry Commission 1998). In this context, the role of the state in legislating, implementing and monitoring change is much more overt, yet governments have been reluctant to embrace such interventionist strategies on the ground that they are both expensive to administer and inconsistent with the new 'enabling' role that advanced liberal states should adopt. Instead, state agencies are increasingly looking to the market as a way of encouraging farmers to use their resources more efficiently. Will environmental problems be 'solved' through market-based options when the market has been implicated, in previous decades, in creating the conditions for resource exploitation?

Social scientists are understandably sceptical of the ability of such reforms to address, let alone reverse, the widespread environmental degradation presently being caused by productivist agriculture. Authors such as James O'Connor (1994: 172) propose a form of 'ecological socialism': 'a society that pays close attention to ecology along with the needs of human beings in their daily life, as well as to feminist issues, antiracism, and issues of social justice and equality more generally'.

Since these goals are invariably long term – and, indeed, may never be realised – other, medium-term strategies are also being promoted that portend a fundamental rethinking of the role of the 'farmer' and the withdrawal of ecologically damaged or significant land from agricultural production. In doing this, farmers would be paid for 'ecosystem services' – land conservation and regeneration services rendered for the benefit of all. Rather than being 'producers' in a market economy, they would be provided with a yearly income to act as caretakers or conservationists by returning land to native forests and vegetation. While stewardship payments are currently in operation in both Australia (The Landmark Project 2002) and the United Kingdom (Clark & Murdoch 1997), both the sum of the compensation offered and the number of farmers involved are too small to have anything other than piecemeal, localised impacts. Effective solutions, therefore, are still years away – years during which environmental degradation will continue.

The alternatives: organic production systems

If the presence of farm-derived environmental pollutants can be attributed to the entrenched agri-chemical 'high-tech' regime, would a reduction in those inputs lead to better environmental outcomes? The answer is a qualified 'yes'. Farmers are embarking on such monitoring systems through integrated pest management (the close evaluation of pest infestations and the sparing use of chemicals) and by adopting quality assurance systems at the insistence of food processors and retailers (Lockie 1998). They are also employing 'conservation farming' techniques which, while they require chemicals to poison weeds, allow producers to reduce or eliminate ploughing, thereby minimising the loss of topsoil. There is also a very strong consumer demand worldwide for products that are free of agri-chemicals and do not contain genetically modified organisms (GMOs) (see Lyons et al. 2004a, 2004b).

With a 25 per cent yearly growth in the rate of consumption of its products, the organic food sector is the fastest growing of any food sector in the world. From a very low base, the organic sector is expected to occupy some 15 per cent of world market share by 2005, with a prediction of sales of some $US100 billion by the year 2010 (Segger 1997). At present the United States and Japan are the largest markets, with Europe close behind. Importantly, the great bulk of organic products sold in the marketplace are not being grown and traded by small-scale 'alternative' farmers, but by large globally oriented firms. Many commercial farmers are in the process of converting to organic production to take advantage of price premiums, and corporations have seized on organics to achieve profits in what are viewed as 'healthy' and 'clean and green' products (Campbell & Coombes 1999). While evidence indicates that organic foods have health benefits for consumers (Heaton 2001), organic farming also produces greater environmental benefits than productivist farming methods (Allen & Kovach 2000). While this does not mean that all organic systems are sustainable production systems (in any system soil nutrients can be depleted if organic matter is not returned fast enough), it is clear that certified organic growers, who must abide by stringent growing requirements to receive accreditation, have the capacity to move farming systems closer to sustainability than conventional producers (see discussions in Altieri 1998). While some writers have argued that organics is simply a 'yuppie fad' and would be unable to feed a starving world (Avery 1995), others have pointed out that the reason people starve is not the lack of food worldwide, but the political, financial and institutional barriers to the better distribution of foods (McMichael 2003). Hence it is likely that organics will have an increasingly important role to play in the 'greening' of the food industry over the next few decades (Lyons et al. 2004b).

FUTURE DIRECTIONS

As this chapter has identified, one of the key issues facing contemporary agriculture is how to continue to provide food and fibre to satisfy growing world demand, yet to do so in a sustainable manner. If productivist agriculture is unable to achieve this, what might take its place? A number of writers have identified what they consider to be the emergence of post-productivist options for rural areas in the advanced countries. Marsden (2003) has proposed that rural space is now becoming a site for consumption as much as production. With urban populations demanding healthy, 'clean and green' foods, and with urban and green groups wanting to preserve the environment, there is a new dynamic in rural areas – one that marginalises conventional agriculture by ensuring that it is simply one of a number of land-use options. With the power of the farmer diminished, and with new actors (such as tourists, recreationalists, indigenous groups and environmentalists) becoming increasingly influential, the shape of farming is thereby modified to match a new ethos that stresses environmental integrity, biodiversity, rural amenity and the symbolic qualities of the countryside (Argent 2002; Marsden 2003). Going beyond post-productivism, writers such as Marsden (2003) have identified another emerging trend – that of rural development, or what has been described in Australia as 'sustainable regional development' (Lawrence 1998; Dore & Woodhill 1999;

Gray & Lawrence 2001). The new dynamic is one which recognises the importance of thinking about society, economy and environment without privileging one over the other two (see Pritchard et al. 2003). While the 'triple bottom line' has become something of a cliché, it is nevertheless a strong reminder that sustainability will only be achieved when market and civil relations are transformed so as to provide policy support for a fundamentally different form of rural production: one that concentrates on long-term sustainable livelihoods for all citizens (Almas & Lawrence 2003; Lawrence 2003; Marsden 2003).

Despite identification of some of the new elements of post-productivism, there are many writers who remain critical of the notion that a post-productivist future is nigh (Wilson 2001; Argent 2002; Lockie et al. forthcoming 2004). Rather than viewing the change as a full transition from one state to another, perhaps it is best to view the changes according to the significance of production, consumption or protection in specified geographical areas. The important insight of Australian geographer John Holmes (2002) is that there are now multiple landscape types emerging in Australia. Yet contemporary rural landscapes are still dominated by production methods that continue to be environmentally damaging. For sociologists, the future research agenda appears to be one of exposing the problematic assumptions underpinning many policy solutions posed by state authorities and various scientific experts; identifying the various 'drivers' that keep contemporary farming in an unsustainable state; and finally, establishing what combination of incentives and penalties might best move rural areas towards sustainable regional development.

Productivist agriculture is causing major environmental degradation throughout the Western world, along with those areas of developing countries that have embraced the practices of 'high-tech' farming. As key facilitators of this process, state agencies throughout the world have done little to address this degradation apart from occasional reforms that do little to challenge the hegemony of the capitalist system. At present there are two identifiable trajectories within productivist agriculture. The first is the replacement of some of the more environmentally unsound practices with new ones – such as zero tillage, integrated pest management and the adoption of industry 'best management' practices. The second is the use of ever more advanced technologies, including GMOs, to stimulate efficiency and productivity gains (see Norton 2001; Hindmarsh & Lawrence 2004). Sociologists have argued, however, that neither of these trajectories will address the fundamental contradictions of productivism since both leave producers on a technological treadmill. In contrast, organic production, which opposes productivist agriculture, particularly GMOs, constitutes a growing segment of the world food market. Some large corporations, recognising a discernible trend towards 'greening' within the food industry, are also embracing organics, and are helping to stimulate the growth of an organics industry worldwide. Certified organic production is, *prima facie*, a more sustainable system of production than high-tech farming and will grow in significance as consumers demand 'clean and green' foods, and as food safety issues become more prominent. At this time, however, organics poses little challenge to productivist agriculture.

Another question, then, is this: can productivism be replaced with new, more environmentally sound, post-productivist approaches? While we are witnessing the reconfiguration of rural space, there is only partial evidence that productivism is being confronted and replaced. What seems to be occurring is that productivist agriculture is being modified in ways described above but still retains its place as one of a number of different landscape forms. Post-productivist options might alleviate some of the symptoms of unsustainability but will not address its root causes. Thus it would appear that one way forward in developing a truly sustainable farming system would be to embrace sustainable regional development, the core of which is a triple bottom line approach to decisions about resource use, economic development and social progress. To date, however, there has been limited evidence to suggest that sustainable regional development is high on the agenda of governments, or that any fundamental change is imminent.

Discussion Questions

1. Is productivist agriculture 'inevitable'? Does it provide the most likely future for the world food economy?
2. Is 'sustainable capitalism' a contradiction in terms?
3. How might transnational food/agribusinesss firms undermine or contribute to a more environmentally sustainable agriculture?
4. Genetically modified organisms are viewed by many scientists and government officials as the potential 'saviour' of modern agriculture. Why?
5. Are farmers to be blamed for failing to adopt sustainable land management practices? What theories or evidence can we draw on to show that this is not the case?
6. Should we be paying farmers for preserving trees and other vegetation? If 'yes', who should bear this cost, and why? If not, how will we ensure that biodiversity is protected?

7. Is organic farming part of the past – or part of the future?

8. What are some of the criticisms that sociologists have made of existing policy approaches to addressing environmental degradation caused by agriculture?

9. Think about the sort of world you will live in in 2020. What will agriculture be like? Will the world's farming systems be 'sustainable'? If 'yes', how will this have come about?

10. Give one example of post-productivism occurring in an area of the world with which you are familiar. Do you believe this will lead to a new trajectory for agriculture in that region?

Glossary of Terms

Genetically modified organisms: biological entities that have been created through alteration of the genetic makeup of cells, usually by the insertion, removal, or manipulation of individual genes.

Productivism: the pursuit of higher profits in farming and grazing by increasing efficiency and productivity. This is achieved by employing 'high-tech' inputs such as hybrid (or genetically modified) seeds, and chemical fertilisers and pesticides, in combination with labour-saving machinery and modern management techniques.

Post-productivism: the marginalisation of productivist farming and grazing as rural space becomes a site both for consumption and environmental protection.

Sustainable development: that form of development that improves the total quality of life, both now and in the future, in a manner that maintains the ecological processes on which life depends.

Treadmill of production: also referred to as the 'technological treadmill'. The continuing use of increasingly potent chemicals and fertilisers (and other technologies) to secure productivity and efficiency gains in the agricultural and grazing industries. The 'treadmill' is an outcome of the adoption of a productivist approach to farming.

References

Allen, P., and M. Kovach 2000, 'The capitalist composition of organic: the potential of markets in fulfilling the promise of organic agriculture', *Agriculture and Human Values* 17: 221–232.

Almas, R., and G. Lawrence (eds) 2003, *Globalisation, Localisation and Sustainable Livelihoods*, Aldershot, UK: Ashgate, pp. 189–203.

Altieri, M. 1998, 'Ecological impacts of industrial agriculture and the possibilities for truly sustainable farming', *Monthly Review* 50(3): 60–71.

Argent, N. 2002, 'From pillar to post? in search of the post-productivist countryside in Australia', *Australian Geographer* 33(1): 97–114.

Avery, D. 1995, *Saving the Planet with Pesticides and Plastic: The environmental triumph of high-yield farming*, Indianapolis IN: Hudson Institute.

Bonanno, A., L. Busch, W. Friedland, L. Gouveia and E. Mingione (eds) 1994, *From Columbus to ConAgra: The globalisation of agriculture and food*, University Press of Kansas.
Burch, D., and R. Rickson 2001, 'Industrialised agriculture: agribusiness, input dependency and vertical integration'. In Lockie and Bourke, *Rurality Bites*, pp. 165–77.
Buttel, F. 2001, 'Environmental sociology and the sociology of natural resources: strategies for synthesis and cross-fertilisation'. In Lawrence et al., *Environment, Society and Natural Resources*, pp. 19–37.
Campbell, H., and B. Coombes 1999, 'Green protectionism and organic food exporting from New Zealand: crisis experiments in the breakdown of Fordist trade and agricultural policies', *Rural Sociology* 64(2): 302–19.
Cary, J., T. Webb and N. Barr 2002, *Understanding Landholders' Capacity to Change to Sustainable Practices*, Canberra: Bureau of Rural Sciences, Agriculture, Fisheries, Forestry – Australia.
Clark, J., and J. Murdoch 1997, 'Local knowledge and the precarious extension of scientific networks: a reflection on three case studies', *Sociologia Ruralis* 37(1): 38–60.
Davis, J. 1980, 'Capitalist agricultural development and the exploitation of the propertied laborer'. In F. Buttel and H. Newby (eds) *The Rural Sociology of the Advanced Societies*, Montclair NJ: Allenheld, Osmun.
Dore, J., and J. Woodhill 1999, *Sustainable Regional Development: Executive summary of the final report*, Canberra: Greening Australia.
Foster, J., and F. Magdoff 1998, 'Liebig, Marx, and the depletion of soil fertility: relevance for today's agriculture', *Monthly Review* 50(3): 32–45.
Gray, I., and G. Lawrence 2001, *A Future for Regional Australia: Escaping global misfortune*, Cambridge University Press.
Heaton, S. 2001, *Organic Farming, Food Quality and Human Health: A review of the evidence*, Bristol, UK: UK Soil Association.
Heffernan, W. 1998, 'Agriculture and monopoly capital', *Monthly Review* 50(3): 46–59.
Herbert-Cheshire, L. 2000, 'Contemporary strategies for rural community development in Australia. A governmentality perspective', *Journal of Rural Studies* 16(2): 203–15.
Herbert-Cheshire, L. 2001, '"Changing people to change things": building capacity for self-help in natural resource management. A governmentality perspective'. In Lawrence et al., *Environment, Society and Natural Resources*, pp. 270–82.
Higgins, V. 2002, 'Self-reliant citizens and targeted populations: the case of Australian agriculture in the 1990s', *Arena* 19: 161–77.
Hindmarsh, R., and G. Lawrence (eds) 2004, *Recoding Nature: Critical perspectives on genetic engineering*, Sydney: UNSW Press.
Holmes, J. 2002, 'Diversity and change in Australia's rangelands: a post-productivist transition with a difference?', *Transactions of the Institute of British Geographers* 27: 362–84.
Ilbery, B., and I. Bowler 1998, 'From Agricultural Productivism to Post-productivism'. In B. Ilbery (ed.) *The Geography of Rural Change*, Harlow, UK: Addison-Wesley-Longman, pp. 57–84.

Industry Commission 1998, A Full Repairing Lease: Inquiry into ecologically sustainable land management. Report No. 60, 27 January, Canberra: Commonwealth of Australia.

Landmark Project, The 2002, 'How to encourage sustainable land use in dryland regions of the Murray-Darling Basin'. Policy Discussion Paper. http://www. landmark.mdbc.gov.au/policy/policy1.htm.

Lawrence, G. 1987, Capitalism and the Countryside: The rural crisis in Australia, Sydney: Pluto Press.

Lawrence, G. 1998, 'The Institute for Sustainable Regional Development'. In J. Grimes, G. Lawrence and D. Stehlik (eds) Sustainable Futures: Towards a catchment management strategy for the Central Queensland Region, Rockhampton, Qld: Institute for Sustainable Regional Development, CQU, pp. 6–8.

Lawrence, G. 2003, 'Sustainable regional development: recovering lost ground'. In Pritchard et al., Social Dimensions of the Triple Bottom Line, pp. 157–70.

Lawrence, G., V. Higgins and S. Lockie (eds) 2001, Environment, Society and Natural Resources: Theoretical perspectives from Australia and the Americas, Cheltenham, UK: Edward Elgar

Lockie, S. 1998, 'Environmental and social risks, and the construction of "best practice" in Australian agriculture', Agriculture and Human Values 15(3): 243–52.

Lockie, S. 2000, 'Environmental governance and legitimation: state–community interactions and agricultural land degradation in Australia', Capitalism, Nature, Socialism 11(2): 41–58.

Lockie, S. 2001, 'Agriculture and environment'. In Lockie and Bourke, Rurality Bites, pp. 229–42.

Lockie, S., and L. Bourke (eds) 2001, Rurality Bites: The social and environmental transformation of rural Australia, Sydney: Pluto Press.

Lockie, S., G. Lawrence and L. Cheshire 2004 forthcoming, 'Nature, culture and globalisation: reconfiguring rural resource governance'. In P. Cloke, T. Marsden and P. Mooney (eds) Handbook of Rural Studies, London: Sage.

Lockie, S., and F. Vanclay (eds) 1997, Critical Landcare, Wagga Wagga, NSW: Centre for Rural Social Research, Charles Sturt University.

Lyons, K., S. Lockie and G. Lawrence 2004a forthcoming, 'Organics, biotechnology and food: Australian consumer views'. In R. Hindmarsh and G. Lawrence (eds) Recoding Nature: Critical perspectives on genetic engineering, Sydney: UNSW Press.

Lyons, K., D. Burch, G. Lawrence and S. Lockie 2004b forthcoming, 'Contrasting paths of corporate greening in antipodean agriculture: organic and green production'. In K. Jansen and S. Vellema (eds) Agribusiness and Society: Corporate responses to environmentalism, market opportunities and public regulation, London: Zed Books.

McMichael, P. 2003, 'The power of food'. In Almas and Lawrence, Globalisation, Localisation and Sustainable Livelihoods, pp. 69–85.

Magdoff, H., F. Buttel and J. Foster (eds) 1998, Hungry for Profit: Agriculture, Food and Ecology, Special Edition, Monthly Review 50(3).

Marsden, T. 2003, The Condition of Rural Sustainability, Assen, Netherlands: Royal Van Gorcum.

Martin, P. 1997, 'The constitution of power in landcare: a post-structuralist perspective with modernist undertones'. In Lockie and Vanclay, *Critical Landcare*, pp. 45–56.

Martinez-Alier, J. 1999, 'The socio-ecological embeddedness of economic activity: the emergence of a transdisciplinary field'. In E. Becker and T. Jahn (eds) *Sustainability and the Social Sciences: A cross-disciplinary approach to integrating environmental considerations into theoretical reorientation*, Paris: UNESCO, pp. 112–39.

Mooney, P. 1988, *My Own Boss?* Boulder CO: Westview.

National Farmers' Federation and Australian Conservation Foundation 2000, *A National Scenario for Strategic Investment*. http://www.nff.org.au/rtc/5point.htm.

Norton, J. 2001, 'Biotechnology to the rescue? Can genetic engineering secure a sustainable future for Australian agriculture?' In Lockie and Bourke, *Rurality Bites*, pp. 270–83.

O'Connor, J. 1994, 'Is sustainable capitalism possible?' In O'Connor, *Is Capitalism Sustainable?* pp. 152–75.

O'Connor, M. 1994a, 'Introduction: Liberate, accumulate – and bust?'. In O'Connor, *Is Capitalism Sustainable?* pp. 1–21.

O'Connor, M. (ed.) 1994b, *Is Capitalism Sustainable? Political economy and the politics of ecology*, New York: Guilford Press.

Pritchard, B., A. Curtis, J. Spriggs and R. Le Heron (eds) 2003, *Social Dimensions of the Triple Bottom Line in Rural Australia*, Canberra: Bureau of Rural Sciences.

Richards, C., G. Lawrence and N. Kelly 2003, 'You can't be green when you are in the red'. 'New Times, New Worlds, New Ideas: Sociology Today and Tomorrow' Conference, Australian Sociological Association. Armidale, NSW, 4–6 December.

Rogers, E. M. 1962, *Diffusion of Innovation*, New York: Free Press.

Schnaiberg, A. 1980, *The Environment*, Oxford: Oxford University Press.

Segger, P. 1997, 'World trade in organic foods: a growing reality'. Paper presented at the Future Agenda for Organic Trade, Fifth IFOAM Conference on Trade in Organic Products, Oxford.

Stoneham, G., M. Eigenraam, A. Ridley and N. Barr 2003, 'The application of sustainability concepts to Australian agriculture', *Australian Journal of Experimental Agriculture* 43: 195–203.

Vanclay, F. 1997, 'The social basis of environmental management in agriculture: a background for understanding landcare'. In Lockie and Vanclay, *Critical Landcare*, pp. 9–27.

Vanclay, F., and G. Lawrence 1995, *The Environmental Imperative: Eco-social concerns for Australian agriculture*, Rockhampton, Qld: CQU Press.

Wilson, G. A. 2001, 'From productivism to post-productivism . . . and back again? Exploring the (un)changed natural and mental landscapes of European agriculture', *Transactions of the Institute of British Geographers* 26: 77–102.

Wright, S. 1990, 'Development theory and community development practice'. In H. Buller and S. Wright (eds) *Rural Development: Problems and practices*, Aldershot, UK: Avebury, pp. 41–63.

CHAPTER FOURTEEN

PATHOLOGICAL ENVIRONMENTS

Peter Curson and Lindie Clark

The environments within which people live and work are not the benign settings we often assume them to be. Some of them are, to differing degrees and for a variety of reasons, potentially pathological. In those parts of the world where the population lacks access to a safe supply of drinking water and adequate waste disposal facilities, for example, diarrhoeal diseases and respiratory tract infections continue to exact a large toll on human health. The persistence of such environmentally related communicable diseases, along with the outbreaks of new ones such as BSE, Ebola, and SARS, makes the claims of the 1960s and 1970s that modern medicine had conquered infectious disease ring increasingly hollow. Meanwhile, the proportion of the global burden of non-communicable disease and injury linked, at least in part, to environmental causes is also on the rise. Work-related injuries and cancers, lung damage from indoor and outdoor air pollution, the rising number of antibiotic-resistant bacterial diseases, and mounting evidence linking neurobehavioural disorders in children to exposure to lead and other toxic chemicals all indicate that the environments in which we live, work, learn, and play can pose serious risks to our health.

What factors account for these trends? Does the increase in environmentally related health risks reflect an underlying increase in the pathogenicity of certain environments, or are specific patterns of human social interaction within those environments instead to blame? Does the rising environmental health toll reflect heightened human awareness of, or perhaps increased vulnerability to, such risks? Why is it that certain groups of people – the young, the old, and the poor, in particular – bear a disproportionate burden of environmentally related injury and disease? This chapter provides an introduction to some of the key contemporary debates in the field of environmental health through a discussion of questions such as these. As well as considering the interaction between humans and potentially pathological agents in the 'natural' or biophysical environment, we also look at how pathological environments can be created 'behind closed doors': in the workplace, in our homes, and even in the very places that we go to for the treatment of disease.

BACKGROUND TO THE ISSUES

Environmental health, as both a field of academic inquiry and an arena for social action, has a long and somewhat turbulent history. Historians of public health argue that just about all societies have implemented some form of collective measure to protect their populations from health risks located in the environment (see Brockington 1979; Rosen 1993). Many such preventive measures have proved effective in the protection of public health, even though the specific aetiology of the types of diseases they were designed to check was not always clearly understood.

The issue of disease causation has in fact been one of the main sources of controversy in the history of environmental health. From Hippocrates' time (c. 430 BC) until the mid-19th century, miasmatic theories, which linked certain environmental conditions and/or specific places with the outbreak of disease (e.g. marshy waters with epidemics of dysentery and malaria), held considerable sway in the Western intellectual tradition, thus boosting the cause of environmental health. Such theories provided the conceptual underpinning for many of the public health reforms (such as the provision of sewerage and safe water supplies) pursued by the 19th-century Sanitary Movement (see Melosi 1980; Coward 1988). But with the development of the germ theory of disease in the latter part of that century, marking the birth of modern medicine and the ascendancy of the biomedical model, miasmatic theories of disease were by and large displaced. As they were, enthusiasm for the cause of environmental health, with which they were closely associated, suffered something of a setback.

Interest in the health–environment nexus and the pathology of certain places, however, was to re-emerge in a number of different guises over the next century. One such instance was the rise of the Geomedizin movement in Germany of the 1930s and 1940s, adherents to which advocated a holistic approach to the study and prevention of disease, viewing it as the result of the interplay of geographical, social, and biological factors (see Weindling 2000). Geomedizin was to become totally corrupted by virtue of its links with Nazi ideas of race, blood, and soil, but a more progressive strand of environmental health was to surface during the period of environmental and related social activism in the advanced capitalist nations dating from the 1960s. The surge of interest in 'green issues' since that time has refocused attention on the human health implications of various forms of environmental degradation (see Hays 1987; Gottlieb 1993; Christoff 1994; Hutton & Connors 1999). Concerns about the toxic effects of exposure to air, water, and soil pollution, overpopulation, and the dangers of the nuclear age dominated the environmental health agenda at first. To these have now been added issues such as the health impacts of global climate change and loss of biodiversity, along with an emphasis on the unequal distribution of environmental risks to health (of which, more below).

KEY DEBATES

Having briefly canvassed some key developments in the history of environmental health, we now move on to consider four issues around which interesting debates are currently taking place in the field. The four issues selected are intended to give a flavour of the scope of environmental health as an academic and social enterprise; they also revolve around problems that are (or we believe should be) of high priority on researchers', policy-makers', and activists' agendas. Those issues are: global climate change and human health; the (re)emergence of infectious disease; pathogenic indoor environments; and the generation of fear and panic in responding to perceived environmental health threats.

Global climate change and human health

Since the early 1990s a number of large-scale ecological problems which have the potential for serious adverse impacts on human health – such as climate change, stratospheric ozone depletion, trans-border pollution, land degradation, and loss of biodiversity – have increasingly attracted the attention of those active in the environmental health field. For some, the question of whether humans will survive the threats posed by such 'global threats of environmental disaster' represents 'the most significant public health question for the twenty-first century' (Baum 2002: 255). Others, sceptical of what they label 'doom and gloom' scenarios, are not so sure. We focus here on the controversies surrounding just one of these large-scale ecological issues, that of climate change, as it is a topic that has attracted a significant level of interest and debate in the environmental health field.

Evidence that the world is gradually warming has become stronger in recent years. Last century, the average annual temperature of the globe increased by about 0.6 ($\pm$ 0.2) degrees and warming has become more marked in the last few decades, as has the incidence of extreme weather events. Debate in the field has generally moved on from the question of whether predictions of global warming are soundly based, the consensus being that they are. The main controversies now centre around the nature and magnitude of the impacts of this phenomenon on human health, and the relative priority they should be accorded vis-à-vis other environmental health concerns.

The consensus on the reality of global warming is not to deny the uncertainty that still exists. Because climate affects us in a myriad often unnoticed ways, and because the actual details of climate change at local and regional level remain far from certain, predicting the human health effects of climate change is very difficult. But that uncertainty should not deter us from considering the possible health consequences of climate change (including whether certain places and populations are particularly vulnerable to its effects), or so a number of environmental health practitioners argue, as such issues are crucial for the design of effective 'countermeasures' that include prevention and early detection (see Haines 1993).

Heeding this call, a considerable body of environmental health literature has sought to predict some of the likely effects of climate change on human health. A recent comprehensive review of this literature prepared from the Intergovernmental Panel on Climate Change (IPCC) summarises the main findings (McMichael & Githeko 2001). *Direct* health impacts of global warming and an increase in extreme weather events are predicted to include an increased risk of: thermal stress-related morbidity and mortality (particularly among older age groups and the urban poor); certain respiratory diseases (including those resulting from increases in the climatically related production of certain air pollutants in large urban areas); and adverse health effects from climate-related disasters (cyclones, floods, famine, etc.), the experience of which is expected to be most severe in developing countries lacking adequate infrastructure. *Indirect* health effects associated with climate change are predicted to result from: changes in the range and transmission of vector-borne and infectious diseases (e.g. an increase in malaria, dengue, and leishmaniasis in higher altitudes and perhaps also in higher latitudes); population displacement and economic disruption as a result of an increase in climate-related disasters and rising sea levels (the latter particularly affecting low-lying island states); and changes in the availability of fresh water and in the productivity of local agricultural systems (which are expected to have net negative effects in the developing world). While some *beneficial* impacts of climate change on human health have also been predicted (e.g. a reduction in winter deaths from pneumonia as a result of warmer winters and fewer cold spells in at least some temperate countries), 'overall, negative impacts are anticipated to outweigh positive health impacts' (McMichael & Githeko 2001).

Such findings, however, have not been universally accepted. Some authors argue that the predictions of unfavourable impacts on health are exceedingly speculative, often based on simplistic thinking and, at least in some cases, dependent on 'soft data' (see Taubes 1997). These critics argue that the gloom and doom scenarios painted by climate modellers are 'great copy', but ignore the capacity of people to adapt their behaviours and lifestyles to cope with climate change. They also maintain that recent outbreaks of vector-borne and other infectious diseases (e.g. the Latin American dengue epidemics of 1994 and 1995), painted by some as harbingers of what is to come, have more to do with breakdowns in public health in many countries than with climate change per se. Focusing undue attention on global warming as a public health issue is problematic, these critics argue, as it 'will distract the public from other priorities' (Gubler, cited in Taubes 1997).

While few would deny that many developing countries are directing less resources to public health, many would take issue with a number of the critics' other assertions and assumptions. The import of the critics' argument also remains unclear: are they suggesting that trying to predict the environmental health consequences of global climate change is a worthless cause? Despite (perhaps because of) the uncertainties involved, are there not compelling reasons to continue such research, not least to raise public and political awareness of the importance of preventing (or at least stalling) global warming? Isn't such research also essential if we are to be in the best possible position to detect and deal with the potential environmental health impacts (e.g. by strengthening public health infrastructure) should they nonetheless eventuate?

Furthermore, and leaving aside its perhaps misplaced optimism, the critics' belief that 'as climate changes, man [sic] changes as well' (Henderson, cited in Taubes 1997) fails to recognise that not all people will be equally affected by the impact of global warming and an increase in extreme weather events, nor will they be equally capable of making the required 'adaptations'. A substantial literature has demonstrated that existing patterns of social and epidemiological vulnerability mean that it is usually the old, the handicapped and disabled, and the disadvantaged who bear the real impact of such events (see Curson 1996). The heatwave which struck Western Europe in August 2003 is a case in point. Perhaps as many as 10 000, mostly elderly people, died in France during a two-week period of extraordinarily high temperatures. Most were elderly people living alone or in severely under-funded and understaffed rest homes. In this case, the general vulnerability of the aged to thermal stress was compounded by social isolation, cutbacks in funding programs and the slowness of the public health system to react. Such problems of vulnerability are multiplied many times over in the developing world.

The (re)emergence of infectious disease

Emerging/re-emerging infections have been on the march since Korean Hemorrhagic Fever was first identified in 1951, but it is only quite recently that the threat they posed to public health has been more fully realised. The failure to appreciate the threat earlier reflects the complacency about infectious disease that prevailed from the late 1960s and the belief that the developed world had entered a new antiseptic age where infections had been totally displaced by lifestyle and behavioural diseases like cardiovascular disease, cancer, and stroke. (Even at the time, this belief ignored the large and persisting health toll exacted by infectious disease on disadvantaged and vulnerable sub-populations in such countries.) In addition, there was a tendency to underestimate the power of the biophysical environment and to believe that microbes were stationary targets against which magic antibiotic bullets could be directed. The upshot was that public health in most developed countries changed its focus to concentrate on degenerative and chronic diseases. It was this decision, one which is now coming back to haunt us, that forms the second 'key controversy' in environmental health discussed here.

The false sense of security into which many in the developed world had been lulled was shaken with the advent of a series of 'new' infections which burst forth after the late 1960s. Within a few years the world had encountered a wide range of new infections, including Lassa Fever, HIV/AIDS, Lyme Disease, Ebola, Rift Valley Syndrome and Legionnaires Disease. At the same time, a wide range of diseases began to develop resistance to antibiotic drugs. This would not have been so much a problem if the large pharmaceutical companies had continued research and development of new anti-microbial drugs to combat resistant strains of bacteria and viruses. Largely, however, many saw more profitable lines of development in the area of providing drugs for degenerative disorders and complaints like arthritis, high cholesterol, asthma, and high blood pressure (see Projan 2003).

Over the last thirty years, approximately forty 'newly' emerged infections have been identified. The majority of these have resulted from long-established zoonoses being brought out of their permanent natural disease reservoirs or provided with a selective advantage by some element of human behaviour and/or environmental change. The emergence and spread of 'new' infections and the re-emergence of long-established ones is driven by a variety of factors (Olshansky et al. 1997). These include: genetic and biological change in disease agents; human-induced changes in the physical environment which impact on the ecology of natural disease systems; human behaviour which individually and collectively can expose people to 'new' or re-emerged infections; international travel and human mobility; the globalisation of trade; and a variety of social, economic and political factors, including the breakdown or lack of public health measures, the growth of mega-cities, and the increase in poverty and marginalisation.

The latter point returns us to our earlier observation that the (false) belief of the postwar period that the developed world was 'done' with infectious disease always ignored the large and persisting health toll exacted by environmentally related communicable disease on certain vulnerable groups within their populations. The case of trachoma, the leading cause of blindness in the world today, illustrates this point. Once a disease of the urban slums, workhouses, and tenements in Europe, the disease was introduced to Australia in the 19th century, and, as had been the case in Europe, became associated with depressed and unsanitary housing allied to local heat, dirt, and flies. By the 20th century, improved hygiene and living conditions led to the disappearance of the disease throughout most of the developed world. In Australia, trachoma disappeared among the white population over a hundred years ago. For sections of Australia's Aboriginal population, however, blinding trachoma remains an important public health problem.

Trachoma remains endemic throughout areas where the living conditions of indigenous communities are depressed and unsanitary, and possibly as many as 30 000 Aboriginal children are afflicted by the disease (Rintoul 2003). Many remote Aboriginal communities still lack access to basic facilities like adequate housing and potable water. Among such communities it would seem that the incidence of trachoma has changed very little over the last twenty or so years, with between 40 and

70 per cent of all children experiencing active trachoma, and in some areas 20 per cent of all elderly people have in-turned eyelashes (Taylor 2001). To eliminate trachoma would require the upgrading of housing, living conditions, and basic services for remote indigenous communities to standards enjoyed by other Australians. So far there would appear to be insufficient political and/or community resolve to achieve this.

Pathological indoor environments

The third controversy highlighted here concerns an area of environmental health which, like environmentally related infectious disease, has been historically underplayed: that of the pathogenic nature of a variety of *indoor* environments. As was the case during the burst of activism around environmental health issues in the mid to late 19th century, interest in the health impacts of pollution reignited by the environmental movement of the 1960s tended to focus on toxins encountered in the ambient (or outdoor) environment. Since then considerable research and socio-political attention has been given to such environmental health threats, for example those posed by hazardous emissions from smokestacks and automobile exhausts. Until fairly recently, the environmental health threats encountered in the indoor environment were by and large ignored. And yet the range and quantum of indoor environmental health threats is large: ranging from the increased risk of disease due to exposure to toxic substances (e.g. fumes, chemicals, fibres, tobacco smoke), physical hazards (e.g. heat, vibration, radiation, noise), and biological agents (e.g. drug-resistant microorganisms), through to injuries and infections linked to unsafe systems of work and inadequate sanitary practice. As this list suggests, the exact nature of indoor threats depends to a large extent on the specific environment in which they are encountered – be it home, work, school, or other public or private institution. This is so not just because of differences in the physical characteristics of such environments, but also because of their differing social and political nature as well. Our discussion of some of these potentially pathogenic indoor environments – hospitals, homes, and workplaces – begins with one that we tend to regard as a refuge from death and disease.

Hospitals

It is an irony in these days of high technology and high-quality health care, that hospitals often belie their true purpose and cause ill health. Hospital or nosocomial infections have been a critical problem affecting the quality of health care and a major source of morbidity and

Table 14.1 *Top ten surgical site infection rates,*
US contributing hospitals, 1992–2002

Operative procedure	Rate per 100 operations
Small bowel	5.06
Organ transplant	4.48
Ventricular shunt	4.17
Limb amputation	3.63
Liver/pancreas	3.14
Other digestive	3.08
Caesarean section	2.83
Other respiratory	2.57
Head and neck	2.33
Coronary bypass with chest incision	2.19

Source: NNIS (2002)

mortality for over 150 years (Gaynes 1997). Most countries have witnessed a cyclical parade of pathogens in hospitals, from a variety of streptococcal infections during the 19th century, to a series of opportunistic drug-resistant infections such as *Staphylococcus aureus*, enterococci and staphylococci in the late 20th century (Weinstein 1998). Hospitals remain wonderful breeding grounds for a wide variety of infectious agents, and today, possibly between 5 and 12 per cent of all hospital admissions will acquire an infection (WHO 2002: 1). In the United States, roughly 5 per cent of all in-hospital patients, or more than 2 million people a year, have the misfortune to acquire an infection in hospital and approximately 100 000 die each year as a result (Expert Working Group 2001). The costs of such circumstances are severe, not only in terms of human suffering, but also in financial terms. Hospital-based infections are thought to account for 2 per cent of total hospital costs in the US (Expert Working Group 2001: 9).

While in Australia the hospital infection rate appears lower than in the United States (3 per cent compared to 5 per cent), the National Hospital Morbidity Database has identified spectacular increases in drug-resistant microorganisms affecting hospital patients over the last six years (see Expert Working Group 2001). These data also indicate that a very high proportion of patients leave hospitals still suffering from an infection acquired in hospital. Data from the US National Surveillance System show the risk associated with different surgical procedures (Table 14.1). In Australia, surgical site infections occur in about 13 per cent of patients having colorectal surgery, 7 per cent of patients having vascular surgery, and 2 per cent having prosthetic orthopaedic surgery (Expert Working Group 2001: 8). Bloodstream infections with bacteria and fungi also continue to cause substantial illness and death in hospital patients. Commonly caused by intravenous catheters, there may be over 7000 cases a year in Australia, with mortality rates of up to 30 per cent.

Why do hospital infections remain an important feature of morbidity and mortality in countries committed to high standards of infection control, public hygiene and strict hospital surveillance? Four major factors would seem responsible. First, the use of anti-microbial drugs in hospitals has undoubtedly encouraged antibiotic resistance and encouraged drug-resistant strains to emerge. The ability of microbes to take evasive action against most established forms of medical treatment is well demonstrated (Courvalin & Davies 2003: 426). Antibiotic-resistant organisms are responsible for perhaps 70 per cent of all hospital-acquired infections. Second, changes in health-care technology, the range of invasive procedures employed with respect to organ and tissue transplants, and the use of catheters, ventilators and prosthetic devices have all produced a conducive entry environment for opportunistic infections. In the case of organ transplants and the use of human donor tissues, reliance on immunosuppressive drugs used to prevent rejection have also weakened the body's immune system and left the host susceptible to infection. Third, many of the patients now admitted to hospitals are older and sicker, have weakened immune systems, are subjected to more multiple invasive procedures, and are confronted by a much wider array of care-givers than ever before. Fourth, hospital overcrowding and lax personal hygiene and infection control seem to play a very important role. Cross-transmission of microorganisms by the hands of health-care workers remains the most probable route of many infections (Pittet 2001).

Homes

Despite the relative lack of political attention it receives, indoor air pollution remains one of the most important hazards to public health throughout the world. This is particularly the case for the world's poorest and most vulnerable people. Approximately 50 per cent of people in developing countries remain dependent on fuels such as coal and biomass for cooking and/or heating. Typically, such fuels are burnt on open stoves without adequate venting or removal of fumes and smoke. Smoke from coal or biomass fires is a lethal mix of carbon monoxide particulates, hydrocarbons, and nitrogen oxides together with a mix of carcinogenic agents. Concentrations of suspended particulates in houses using open fires and coal or biomass fuel sources can exceed international standards by up to a hundred times (Saksensa & Smith 1999). Research suggests that such a mix is linked to high levels of respiratory disease, as well as possibly lung, nasopharyngeal and laryngeal cancers, TB, and possibly low birth weight and infant mortality (Bruce et al. 2000). Women and young children are particularly at risk in such situations, given that they spend a number of hours each day in close proximity to the stove.

There is little doubt that indoor air pollution is inextricably linked to poverty and the inability to access cleaner but more expensive fuels such as kerosene, natural gas, and electricity. What then is the solution? Given that poverty prevents the vast majority of people exposed to such risk from changing to less polluting fuels, the fundamental problem that needs to be addressed is clearly socio-economic. In the interim, moving towards a system of efficient venting measures (such as chimneys to take smoke and fumes outside the dwelling) could reduce indoor air pollution by as much as 80 per cent. Some have argued that developing international minimum standards for clean air within buildings, for example under the auspices of the World Health Organization, could also be part of the solution. Leaving aside the important question of how compliance with such regulations could be assured, others point out that setting standards in and of itself will do little unless they are accompanied by practical measures (e.g. funding assistance for ventilation fit-outs) that enable people to address the hazard at source.

Workplaces

It is estimated that in Australia each year, more than 2700 people die as a result of injuries or diseases sustained in the course of their employment (Driscoll & Mayhew 1999). Of these workplace deaths, around 400–450 are the result of traumatic injuries at work (NOHSC 2000: 22); the other 2300 deaths occur as the result of occupationally acquired diseases (Driscoll & Mayhew 1999), most of which are the result of exposure to hazardous substances such as toxic chemicals and fibres. In addition to these work-related fatalities, each year some 650 000 (or 1 in 12) Australian workers suffer a non-fatal injury or illness. In at least 170 000 of these cases, the injuries or illnesses are severe enough that workers require five or more days off work as a result (NOHSC 1996). Certain workers bear a disproportionate share of this workplace injury and disease burden: 'blue-collar' workers (labourers, plant and machine operators, drivers, and tradespeople) account for nearly three-quarters of all new cases each year, even though they constitute only about a third of the total Australian workforce (Foley 1997).

On the basis of this record, it is difficult to conclude that workers' health is being accorded the sort of social and political priority it deserves. This is also apparent if we compare the stringency of regulatory measures in place to protect workers from health hazards encountered in the occupational environment, with those designed to protect the general public from similar hazards arising in the ambient environment (e.g. via pollution control and public health laws). Although the latter are far from perfect, they appear to be significantly more protective of the public's health than are the occupational health and safety laws of workers. For example, while the current Australian environmental health goal is to reduce the amount of lead in the blood of the general public to below 10μ g/dl (NH&MRC 1993), the national occupational health standard permits blood lead levels in Australian workers up to five times that amount (NOHSC 1994). Furthermore, maximum permissible airborne concentrations of lead are almost 200 times more stringent in the ambient, as compared to the workplace, environment (see Table 14.2, p. 250).

Although one might presume that environmental hazards would be regulated on the basis of the danger they pose to human health – rather than according to whether they are encountered in the ambient or the workplace environment – that is plainly not the case in Australia today. This disturbing situation, which results in the unequal protection of workers' health vis-à-vis that of citizens, is not unique to lead (see Table 14.2), nor is it a peculiarly Australian phenomenon. Differentials in stringency between ambient and workplace environmental regulations are found right across the world, apparently irrespective of a nation's political, economic, or social configuration (Derr et al. 1981: 31). Despite the pervasiveness of the workplace/public place regulatory differential, public discussion or even acknowledgement of unequal protection is rare. Raising public awareness of this issue, and encouraging debate, would be a first step towards addressing the inequalities embedded in such laws.

Psychosocial responses to perceived environmental health threats: the generation of fear and panic

It is interesting to speculate on whether the climate of fear that so often accompanies the outbreak of 'new' epidemics, the threat of bioterrorism, or the labelling of yet another component of our diet as

Table 14.2 *Australian air-quality standards for ambient and workplace environments*

Pollutant	Ambient air-quality goal[a]	Workplace air-quality standard[b]	Ratio [Column (3)/(2)]
Carbon monoxide	9 ppm[c]	30 ppm	3.3
Formaldehyde	0.10 ppm[d]	1 ppm	10
Lead	0.50μg/m^{3e}	150μg/m^3	198[f]
Nitrogen dioxide	0.12 ppm[g]	3 ppm	18.75
Ozone	0.08 ppm[h]	0.10 ppm	1.25
Sulphur Dioxide	0.08 ppm[i]	2 ppm	25

Sources: Compiled from NH&MRC (1982); NOHSC (1995); NEPC (1998).
'ppm' = parts per million
Notes: [a] Except for formaldehyde (see note (d) below), the ambient air-quality goals are those adopted by the National Environment Protection Council (NEPC) in June 1998.
[b] The workplace air-quality standards are those recommended by Worksafe Australia (NOHSC 1995). All are expressed as time weighted averages over an 8-hour working day.
[c] Eight-hour average.
[d] There is currently no NEPC goal for formaldehyde: that adopted by the National Health & Medical Research Council (NH&MRC) in 1982 is shown instead.
[e] For a 24-hour period, averaged over a calendar year.
[f] The occupational health standard for lead is averaged over 8 hours, while the environmental health goal is based on 24-hour exposures averaged over 1 year. The maximum cumulative yearly doses permitted by the relevant authorities in both environments were derived to enable calculation of the ratio (i.e. 36,000μg/m^3/182μg/m^3 = 198).
[g] One-hour average.
[h] Four-hour average.
[i] Calendar day average.

potentially carcinogenic, is more threatening to our health than the disease agents themselves. Sociologists have not paid much attention to the way human emotions like fear influence social life and behaviour, particularly during times of perceived crisis (see Tudor 2003); nonetheless, it is a question that warrants much more academic, social, and political discussion, particularly in an environmental health context. The issue of psychosocial responses to perceived environmental health threats forms the final controversy considered here.

The SARS outbreak of 2003 clearly demonstrates how such events produce value-laden social perceptions and responses, and how the psychosocial dimension of such events often heavily outweighs the demographic impact. SARS also demonstrates how fear can impact on people's lives in terms of their use of public space, their involvement in community life, and their attitudes to 'outsiders'. It was at the personal level that the SARS epidemic had its greatest impact. The epidemic brought back classic human reactions to epidemic disease, including fear, hysteria, panic, flight, scapegoating, rumour-mongering and avoidance, as well as a series of draconian government reactions involving surveillance, isolation, quarantine, cleansing and scavenging. People avoided public spaces like cinemas, restaurants and shops. Masks and gloves and excessive hand washing were seen everywhere, and in some countries the Chinese were accused of spreading the epidemic.

SARS brought to a head the matter of what is 'reasonable' fear in such circumstances and the role of the media. Sandman, arguing from the perspective of 'risk communication', argues that 'appropriate' or 'reasonable' fear is a normal emotional state that leads to public readiness, vigilance, prevention and preparedness strategies, and that fear of things like SARS is fundamental to normal human psychology and an integral part of coping strategy (Sandman & Lanard 2003a, 2003b). But those bearing the brunt of the panic may well hold a different view.

The growth of fear stems partly from the way it is created and universalised by the media, commerce, and partly from from the way it is managed through the political process. For the media, the line between 'newsworthiness' and communication, and sensationalism is often a very fine one. Fear 'sells' and this relates closely to the media's entertainment role to attract audiences and make profits. Such an approach places a premium on visual imagery over words, with appeal to the emotions over objective detached reporting, and the accentuation of drama, conflict, and personal tragedy usually outside any social or historical context. Fear resonates with audiences, and the media respond by personalising crises by delineating victims and scapegoats. Yet several decades of research has not provided us with a definitive statement on whether the media's reports about fear are a 'cause' or an 'effect' of public concerns about fear (Altheide 2002: 24).

Fear of fear and the challenge of how to communicate comparative risk to the public without raising undue alarm may also lead to governments falling back on secrecy, miscommunication and misinformation. This happened, not only in China during the early days of the SARS epidemic, but also in Britain, when politicians, government officials and scientific advisors withheld vital information and stifled open discussion during the BSE epidemic for fear of the public's ability to handle such information (see Coghlan 2000; Editorial 2000). In both cases, it would seem likely that the epidemics might have been curtailed earlier if a different line had been pursued.

FUTURE DIRECTIONS

The field of environmental health is by no means the exclusive province of those wearing stethoscopes, those peering down microscopes, or those penning treatises behind the walls of academe. Environmental health, both historically and in contemporary practice, is as much a social and political concern as it is a professional or academic one. This is one of the most important and recurring themes that emerges from the contemporary debates in environmental health canvassed in this chapter. We trust that the intersectoral nature of the environmental health enterprise will be increasingly recognised and embraced in the future; so too its multidisciplinary and interdisciplinary nature (see Raeburn & Macfarlane 2003).

As to areas for future research and action in the environmental health field, the canvas is large. Throughout this chapter we have identified a number of specific questions that we believe warrant further research, public debate and/or political action in relation to the particular controversies discussed: global climate change and human health; the (re)emergence of infectious disease; pathogenic indoor environments; and the discourses of fear and panic generated in response to perceived environmental health threats. A common thread throughout our discussion has been the way in which environmental health risks impact unequally within and between different places and populations. Certain groups – the young, the old, the poor, and certain racial/ethnic minorities – are particularly vulnerable to environmental health risks, and they bear a disproportionate share of the toll in terms of injury (both physiological and psychosocial), death, and disease. It is vital that future research and action are cognisant of, and seek to redress, these inequalities, not just in relation to the range of issues canvassed in this chapter, but in the environmental health field as a whole.

Discussion Questions

1. Outline some key developments in the history of environmental health.
2. What factors underlie the recent global resurgence in infectious disease?
3. How and why do environmental health problems impact unequally on different population groups?
4. Pick an environmental health issue that has arisen in your local community and identify the particular groups on which it impacted most.
5. Identify the key environmental, behavioural, and social changes accounting for the increase in nosocomial (hospital-related) infections.

6. What are the pros and cons of trying to predict the potential environmental health impacts of global climate change?

7. Design a comprehensive program to address the problem of indoor air pollution in developing countries.

8. In what ways is environmental health a social and political issue?

9. What factors might account for the unequal protection of workers and citizens by occupational and environmental health standards?

10. How can governments and the media inform the public about potential environmental health threats without causing a climate of fear and panic?

Glossary of Terms

Aetiology: the study of the causes of disease.

Antibiotic resistance: the phenomenon whereby disease-causing microorganisms develop ways to survive antibiotics that are meant to kill or weaken them, thereby making many diseases harder to treat.

Biomedical model: the dominant approach to the study and treatment of disease, particularly associated with the rise of modern medicine. Biomedical approaches view illness as a malfunction of the individual body's biological mechanisms; they downplay or ignore the role of social and environmental factors in disease causation.

Communicable disease: diseases transmitted to people as a result of direct or indirect contact with other humans, animals, or environments that carry a disease-causing microorganism (e.g. HIV/AIDS, SARS, and the common cold).

Environmental health: the field of study, practice, and social concern dealing with health and illness that result from interactions between people and the environment.

Germ theory of disease: the theory, developed by Louis Pasteur and Robert Koch in the late 19th century, that diseases are caused by specific microorganisms (or 'germs') entering and infecting the human body.

Miasmatic theories: theories of disease causation which link certain environmental conditions and/or specific places with the outbreak of disease.

Non-communicable disease: illnesses or conditions *not* caused by exposure to an infectious agent (see *Communicable disease*). The three main categories of non-communicable disease are cancer, cardiovascular disease, and injuries.

Nosocomial infections: infections acquired by a patient while, and as a direct result of being, in a hospital.

SARS (Severe Acute Respiratory Syndrome): a viral respiratory illness first reported in February 2003. According to the World Health Organization, over 8000 people became sick with SARS in 2003, of whom nearly 800 died.

Zoonoses: animal diseases, transmittable to humans in certain circumstances.

References

Altheide, D. 2002, *Creating Fear: News and the construction of crisis*, New York: Aldine de Gruyter.

Baum, F. 2002, *The New Public Health*, 2nd edn, Melbourne: Oxford University Press.

Brockington, C. 1979, 'The history of public health'. In W. Hobson (ed.) *Theory and Practice of Public Health*, New York: Oxford University Press, pp. 1–8.

Bruce, N., P.-P. Rogelio and R. Albalak 2000, 'Indoor air pollution in developing countries: a major environmental and public health challenge', *Bulletin of WHO* 78(9): 1078–92.

Christoff, P. 1994, 'Environmental politics'. In J. Brett, J.A. Gillespie and M. Goot (eds) *Developments in Australian Politics*, Melbourne: Macmillan, pp. 348–67.

Coghlan, A. 2000, 'How it went so horribly wrong', *New Scientist* 2263 (4 November): 4–6.

Courvalin, P., and J. Davies 2003, 'Antimicrobials: time to act', *Current Opinion in Microbiology* 6: 425–6.

Coward, D.H. 1988, *Out of Sight: Sydney's environmental history, 1851–1981*, Canberra: Department of Economic History, ANU.

Curson, P. 1996, 'Human health, climate and climate change: an Australian perspective'. In T. Giambelluca and A. Henderson-Sellers (eds) *Climate Change: Developing Southern Hemisphere Perspectives*, Chichester, UK: Wiley, pp. 319–48.

Derr, P., R. Goble, R.E. Kasperson and R.W. Kates 1981, 'Worker/public protection: the double standard', *Environment* 23(7): 6–15, 31–6.

Driscoll, T., and C. Mayhew 1999, 'Extent and cost of occupational injury and illness'. In C. Mayhew and C.L. Peterson (eds) *Occupational Health and Safety in Australia: Industry, public sector and small business*, Sydney: Allen & Unwin, pp. 28–51.

Editorial 2000, 'End of an era', *New Scientist* 2263 (4 November): 3.

Expert Working Group on Australian Infection Control 2001, *National Surveillance of Healthcare Associated Infection in Australia*. Draft Report to Commonwealth Department of Health and Aged Care, Canberra.

Foley, G. 1997, 'Bulletin No. 1. Trends over recent years', *Australian Occupational Health and Safety Statistics*, March. http://www.worksafe.gov.au/worksafe/04/bulls/bull-1.htm.

Gaynes, R. 1997, 'Surveillance of nosocomial infections', *Infection Control and Hospital Epidemiology* 18(7): 475–8.

Gottlieb, R. 1993, *Forcing the Spring: The transformation of the American environmental movement*, Washington DC/Covelo CA: Island Press.

Haines, A. 1993, 'The possible effects of climate change on health'. In E. Chivian, M. McCally, H. Hu and A. Haines (eds) *Critical Condition: Human Health and the environment*, Cambridge MA: MIT Press, pp. 151–70.

Hays, S.P. 1987, *Beauty, Health, and Permanence: Environmental politics in the United States, 1955–1985*, Cambridge University Press.

Hutton, D., and L. Connors 1999, *A History of the Australian Environment Movement*, Cambridge University Press.

McMichael, A., and A. Githeko (eds) 2001, 'Human health'. In Intergovernmental Panel on Climate Change (IPCC) (ed.) *Climate Change 2001: Impacts, adaptation, and vulnerability*, Cambridge University Press, pp. 453–85.

Melosi, M.V. 1980, *Pollution and Reform in American Cities, 1870–1930*, Austin TX/London: University of Texas Press.

NEPC (National Environment Protection Council) 1998, *National Environment Protection Measure for Ambient Air Quality*, Adelaide: NEPC Service Corporation.

NH&MRC (National Health & Medical Research Council) 1982, *Report of the 93rd Session*, Canberra: AGPS.

NH&MRC 1993, *Report of the 115th Session*, Canberra: AGPS.

NNIS (National Nosocomial Infections Surveillance System) 2002, 'NNIS Report: Data Summary from January 1992 to June 2002', *American Journal of Infection Control* 30: 458–75.

NOHSC (National Occupational Health and Safety Commission) 1994, 'National standard for the control of inorganic lead at work' [NOHSC: 1012(1994)]. In *Control of Inorganic Lead at Work*, Canberra: AGPS.

NOHSC 1995, 'National Exposure Standards' [NOHSC:1003(1995)]. In *Exposure Standards for Atmospheric Contaminants in the Occupational Environment*, Canberra: AGPS.

NOHSC 1996, 'Work-related injury and disease: a significant social and economic burden'. Media release, 28 August.

NOHSC 2000, *Data on OHS in Australia: The overall scene*, Sydney: Commonwealth of Australia.

Olshansky, S.J., B. Carnes, R.G. Rogers and L. Smith 1997, 'Infectious diseases: new and ancient threats to world health', *Population Bulletin* 52(2): 2–52.

Pittet, D. 2001, 'Improving adherence to hand hygiene practice: a multidisciplinary approach', *Emerging Infectious Diseases* 7(2): 234–40.

Projan, S. 2003, 'Why is Big Pharma getting out of antibacterial drug discovery?' *Current Opinion in Microbiology* 6: 27–30.

Raeburn, J., and S. Macfarlane 2003, 'Putting the public into public health: towards a more people-centred approach'. In R. Beaglehole (ed.) *Global Public Health: A new era*, Oxford: Oxford University Press, pp. 243–52.

Rintoul, S. 2003, 'Our desert blind spot', *Australian*, 27 August: 9.

Rosen, G. 1993, *A History of Public Health*, 2nd edn, Baltimore MA: Johns Hopkins University Press.

Saksensa, S., and K. Smith 1999, 'Indoor air pollution'. In G. McGranahan & F. Murray (eds) *Health and Air Pollution in Rapidly Developing Countries*, Stockholm: Environment Institute, pp. 111–25.

Sandman, P., and J. Lanard 2003a, 'Fear is spreading faster than SARS – and so it should', *Peter Sandman Risk Communication Web Site*. http:\www.psandman.com.

Sandman, P., and J. Lanard 2003b, 'Fear of fear: the role of fear in preparedness – and why it terrifies officials', *Peter Sandman Risk Communication Web Site*. http:\www.psandman.com.

Taubes, G. 1997, 'Apocalypse not', *Science* 278(5340), 7 November: 1004–6

Taylor, H. 2001, 'Trachoma in Australia', *Medical Journal of Australia* 175: 371–2.

Tudor, A. 2003, 'A (macro) sociology of fear?' *Sociological Review* 51(2): 238–56.

Weindling, P. 2000, *Epidemics and Genocide in Eastern Europe, 1890–1945*, Oxford: Oxford University Press.

Weinstein, R. 1998, 'Nosocomial Infection Update', *Emerging Infectious Diseases* 4(3): 416–20.

WHO (World Health Organization) 2002, *Prevention of Hospital-Acquired Infections: A practical guide*, Geneva: WHO Department of Communicable Disease, Surveillance and Response.

CHAPTER FIFTEEN

ASSESSING THE SOCIAL CONSEQUENCES OF PLANNED INTERVENTIONS

Frank Vanclay

Communities experience continual change. Much of that change is deliberate, the result of planned interventions (policies, plans, programs or projects) that have been initiated and/or implemented by government, at local, state and national levels, by the private sector, or by community groups. While planned interventions are undertaken to achieve desired objectives, there are always unplanned adverse impacts on the environment and on people. In order to consider the appropriateness of a specific intervention, such as whether regulatory approval should be given for a project, and/or what conditions should be imposed, assessment of the likely consequences, both positive and negative, is required.

Projects have many obvious negative consequences on people who live near the site. They also have less obvious impacts that only emerge over time or with careful analysis. Policies, plans and programs also have adverse impacts. Consideration of impacts in the design phase can lead to a reduction in the adverse consequences and to an increase in the benefits of the planned intervention. Planning should be a dynamic process that uses adaptive management principles to iteratively consider the outcomes (positive and negative) of the intervention.

Social Impact Assessment (SIA) is the process of 'analysing, monitoring and managing the social consequences of development' (IAIA 2003: 1). SIA, in its ideal form at least, is a powerful approach that can lead to many benefits. Communities benefit by having more say in decisions; they become revitalised through participation; social capital is built; harmful impacts are avoided; and benefits for communities are maximised. Government agencies benefit by having better information. The private sector benefits by having improved relations with local communities, workforces, and important stakeholders; costly mistakes are avoided; siting decisions are improved; and the risk of future compensation payouts (or fines) is reduced.

While SIA has existed for over thirty years, it has not been fully embraced by society. Governments have not fully committed to it and have not empowered it by withholding resources and endorsement. The corporate sector has been wary of it. Sections of the community have felt betrayed by it because some bad projects have proceeded apparently legitimised by inadequate SIA processes. Despite this, SIA offers real promise as a process of improving planned interventions. In Australia and many other countries, SIA is required for all major projects and for new policy initiatives. Local government planning processes require consideration of social issues, even for small developments. In addition to being required for major projects like the Chevron Papua New Guinea–Queensland gas pipeline, or the Alice Springs–Darwin railway line, SIAs have been used to inform policy initiatives such as the Regional Forest Agreement process, and water-sharing plans in the Living Murray Initiative. The Council of Australian Governments requires all regulatory proposals to undergo impact assessment including 'of the relative social costs and benefits' (COAG 1997: 7). This chapter provides an overview of SIA, focusing on SIA as a field of research and practice. It discusses the key debates in that research field.

BACKGROUND TO THE ISSUES

SIA at the macro level is a paradigm, a field of research and practice, a discourse, or sub-discipline. The SIA field is interested in the processes of analysing, monitoring and managing the social consequences of planned interventions, and by logical extension, the social dimensions of development generally. Planned interventions refer to a policy, program, plan or project that is devised in order to achieve a specific goal. As a field of research and practice, SIA consists of a body of knowledge about theory and methods, a stock of tools, accumulated practical experience, insight, a set of case studies, and shared professional values. These have been codified in the *International Principles for Social Impact Assessment* (IAIA 2003) and in textbooks. SIA is an interdisciplinary social science incorporating sociology, anthropology, demography, development studies, gender studies, human and cultural geography, political science, psychology, social research methods and environmental law.

SIA is also a methodology, framework or approach for analysing, monitoring and managing the social consequences of planned interventions. SIA practitioners use this approach to contribute to

development. They work with communities to achieve better outcomes for communities. They work with development agencies and private sector companies to design better projects and policies, and they
work with regulatory agencies to provide information for the approval
process.

In its original formulation, but now only in its narrowest conceptualisation, SIA is merely a tool, technique or instrument for predicting social impacts as part of Environmental Impact Assessment (EIA).
When so viewed, SIA is a tool used to predict the social impacts that
arise from some planned course of action, usually a proposed project
such as a new mine, dam, highway or factory.

EIA 'is a process of identifying and predicting the potential environmental impacts . . . of proposed actions, policies, programmes and
projects, and communicating this information to decision makers
before they make their decisions on the proposed actions' (Harvey
1998: 2). EIA emerged in the early 1970s with the introduction of the
US National Environment Policy Act of 1969 (NEPA). SIA emerged
alongside EIA and as a consequence of NEPA – and perhaps as a result
of the failure of NEPA and EIA to adequately address social issues.
More detailed accounts of the history of SIA are provided by Burdge &
Vanclay (1995), Becker (1997) and Vanclay (1999).

The NEPA process has been emulated and expanded around the
world. In Australia, EIA requirements were introduced in New South
Wales and Queensland in 1973, with the Commonwealth adopting the
Environmental Protection (Impact of Proposals) Act in 1974. European
countries have been late to adopt EIA partly because their planning
processes were more effective and thus there was less need for such
legislation.

A problem with EIA is an ambiguity about what 'environmental
impacts' refer to, and specifically, whether the social is included. In a
review of EIA texts, Lockie (2001: 277) suggests that many give 'no
more than passing reference' to social impacts. SIA, therefore, is impact
assessment that focuses specifically on the social considerations, rather
than on biophysical (environmental) issues. Impact assessment is a generic term that can mean either an integrated approach, or the composite of all forms of impact assessment such as EIA, SIA, health impact
assessment and so on. But SIA is more than a social version of EIA.

In contrast to EIA, SIA has developed as more than a technique
or step, and is more than the prediction of impacts. There are many
reasons for this. First, it was always considered in SIA that impact
mitigation was more important than impact prediction. The social

conscience of SIA professionals means that they cannot simply advise on the negative impacts but need to be involved in their amelioration and in improving the design of the planned intervention. It was obvious to SIA professionals that much could be done, often with little effort, to reduce social harm and increase benefits. SIA professionals consider that they have an ethical responsibility and a professional duty of care in this regard.

Second, there is an understanding in SIA that often the biggest impact is the fear and uncertainty generated by an intervention. Managing implementation by involving the public is central to harm minimisation. Unlike EIA, which sees public *consultation* as a form of data collection, SIA sees public *participation* as a necessary part of mitigation, a desirable part of project design, and as a human right.

Third, while there is a view in EIA that good science can reasonably predict the impacts that are likely to occur, in SIA there is acknowledgement that social impacts are difficult to predict. Social impacts are interconnected with environmental impacts (Slootweg et al. 2001) and lead to higher-order impacts. Case experience has shown that the higher-order impacts are often more severe than first-order impacts. It is more important to have an ongoing process of management, monitoring and mitigation of impacts than to simply worry about their prediction.

Fourth, no matter how effective predictions might be, there will always be unexpected impacts. With every planned intervention there needs to be an ongoing monitoring process to identify the unexpected impacts when they occur and to respond adaptively to those impacts.

Finally, in SIA, it is expected that planned interventions create benefits as well as harm. There is a role for SIA in enhancing benefits and not just in minimising adverse consequences. By contrast, the culture of EIA has been about harm minimisation rather than benefit maximisation.

For these reasons, SIA is a process of public participation and mitigation, monitoring and management, rather than a task of prediction. Egre & Senecal (2003: 224) summarising their personal experience with three projects, the Three Gorges Dam in China, the Ilisu project in Turkey, and the Urra 1 project in Colombia, conclude:

> Lessons drawn from the three cases also show that pre-project SIAs only amount to a starting point. SIAs will not prove effective, in regard to the maintenance or improvement of socio-economic conditions, if they are not followed up by the implementation of equitable mitigation or resettlement plans developed with the participation of all stakeholders, supported by sufficient budgetary allocations and regularly reviewed

according to the results of monitoring or follow-up programs. Another important (and parallel) condition, which is not often met, is the sustained involvement of SIA practitioners, ideally with the same nucleus of specialists, from feasibility studies to the implementation of mitigation or resettlement plans.

Critiques of EIA focus on its impotence. EIA is seen by developers as a bureaucratic hurdle, and by the community as something that legitimates bad development. Little attention is placed on contributing to design, mitigating side-effects, monitoring for unexpected consequences, or ensuring that projects are suitable paths for development.

In advanced capitalist nations, EIA and SIA are part of the legal process of protecting individual and communal property rights. Clear statements of impacts are required to ensure that these rights are not transgressed. 'Property rights', in its legal-philosophical meaning, include anything that pertains to 'ownership' and access rights, not only real estate (property) but also environmental amenity values and liveability issues such as noise, odour and other forms of pollution, as well as human rights. The proponent (developer) is seen as the potential transgressor of these rights. EIA and SIA are the legal mechanisms by which those rights are assessed to determine if a violation will occur. The EIA process is meant to protect global property rights and shared resources (the public good), while the SIA process is meant to protect individual property rights. In this understanding, EIA and SIA are limited to considering the negative impacts of projects, and have no role in ensuring that good development occurs.

In other contexts, such as in developing countries, there is less need to be concerned about negative impacts on small groups of individuals or on individual property rights. Rather, there needs to be greater concern with maximising social utility and development potential, and protecting and reconstructing sustainable livelihoods while ensuring that development is equitable, sustainable, and generally acceptable to the community. SIA is a philosophy about development and democracy that considers the

- pathologies of development (i.e. harmful impacts),
- goals of development (clarifying what is appropriate development, improving quality of life), and
- processes of development (e.g. participation, building social capital) (Vanclay 2003a: 2).

SIA is multidimensional. An SIA undertaken on behalf of a multinational corporation as part of internal procedures is different to an SIA undertaken by a consultant in compliance with regulatory or funding agency requirements, or an SIA undertaken by a development agency interested in ensuring the best outcomes of their development assistance. These are different from an SIA undertaken by university staff or students on behalf of the local community, or an SIA undertaken by the local community itself. There are many different models about how SIA should be implemented, and different views about the order in which the various steps in each model should be undertaken. Nevertheless, there is consensus among SIA professionals that effective SIA

- participates in the environmental design of the planned intervention;
- identifies interested and affected peoples;
- facilitates and coordinates the participation of stakeholders;
- documents and analyses the local historical setting of the planned intervention so as to be able to interpret responses to the intervention, and to assess cumulative impacts;
- collects baseline data (social profiling) to allow evaluation and audit of the impact assessment process and the planned intervention itself;
- gives a rich picture of the local cultural context, and develops an understanding of local community values, particularly how they relate to the planned intervention;
- identifies and describes the activities which are likely to cause impacts (scoping);
- predicts (or analyses) likely impacts and how different stakeholders are likely to respond;
- assists evaluating and selecting alternatives (including a no-development option);
- assists in site selection;
- recommends mitigation measures;
- assists in the valuation process and provides suggestions about compensation (non-financial as well as financial);
- describes potential conflicts between stakeholders and advises on resolution processes;
- develops coping strategies for dealing with residual or non-mitigatable impacts;
- contributes to skill development and capacity-building in the community;

- advises on appropriate institutional and coordination arrangements for all parties;
- assists in devising and implementing monitoring and management programs. (IAIA 2003: 4)

SIA can be applied at the project level, at the plan or program level, and at the policy level. Any planned intervention has a lifecycle of four stages: planning, construction (or implementation), operation, and decommissioning (phasing out). Each of these phases has its unique impacts and issues to consider and needs to be considered separately. Because of the potential presence of fear, anxiety and speculation, social impacts can also occur not only at the level of a plan, but even when an intervention is just an idea.

KEY DEBATES

The definition of 'the environment' debate

A major dilemma in EIA is what 'the environment' means. For most writers, and Harvey (1998: 2) is typical, environmental impact means 'bio-geophysical, socio-economic and cultural'. This holistic notion of EIA is present in some regulatory contexts. In Australia, the Federal Government's *Environment Protection and Biodiversity Conservation Act 1999*, s. 528, defines the environment as including:

(a) ecosystems and their constituent parts, including people and communities; and

(b) natural and physical resources; and

(c) the qualities and characteristics of locations, places and areas; and

(d) the social, economic and cultural aspects of a thing mentioned in paragraph (a), (b) or (c).

Other Australian legislation, such as *Local Government (Planning and Environment) Act 1990* of the State of Queensland, s. 1.4, uses a similar definition but with sub-sections (c) and (d) being slightly reworded to emphasise the social:

(c) the qualities and characteristics of locations, places and areas, however large or small, that contribute to their biological diversity and integrity, intrinsic or attributed scientific value or interest, amenity, harmony, and sense of community; and

(d) the social, economic, aesthetic, and cultural conditions that affect, or are affected by, things mentioned in paragraphs (a) to (c).

In Australia, therefore, at least until recently, in legislation if not in practice, the term 'environment' was broadly understood. Unfortunately, in some subsequent legislation, for example the *Gene Technology Act 2000*, s. 10, this definition has been condensed. The Explanatory Memorandum relating to this Act (page 48) states very bluntly: 'It is intended that the definition of environment include all animals (including insects, fish and mammals), plants, soils and ecosystems (both aquatic and terrestrial)'. In other words, by its omission the social is no longer included in the definition, and not considered by the Office of the Gene Technology Regulator. It is not clear whether this signals an intended reduction in focus, or merely brings legislation into line with what was initially intended. Nevertheless, for those who would like to see an expansion of social considerations, this change is disappointing. Around the world, considerable confusion exists as to whether 'the environment' includes 'the social' or not. Regrettably, even where there is an intention for the environment to be all-inclusive, too often the social considerations are subordinate to biophysical ones.

The definition of 'social impact assessment' debate

There has been a controversy about the precise definition of SIA. While the *International Principles for Social Impact Assessment* is to some extent a codification of the discipline, there is still a considerable view that sees SIA as being constrained to the prediction of negative impacts within a regulatory framework.

> We define social impact assessment in terms of efforts to assess or estimate, in advance, the social consequences that are likely to follow from specific policy actions (including programs and the adoption of new policies), and specific government actions (including buildings, large projects and leasing large tracts of land for resource extraction), particularly in the context of the U.S. National Environmental Policy Act of 1969. (Interorganizational Committee 1994: 108; see also Interorganizational Committee 2003)

This definition can be contrasted with the definition adopted as part of the *International Principles for Social Impact Assessment*.

> Social Impact Assessment includes the processes of analysing, monitoring and managing the intended and unintended social consequences, both positive and negative, of planned interventions (policies, programs, plans, projects) and any social change processes invoked by those interventions. Its primary purpose is to bring about a more sustainable and equitable biophysical and human environment. (IAIA 2003: 2)

The original understanding of SIA is inherently limiting on a number of counts. It presumes an adversarial regulatory system. It denies that assessment might be carried out internally by a corporation, by government, or even by a community itself independent of a regulatory process. The assessment of impacts of past developments is excluded. There is no role for the management, mitigation and monitoring of impacts, or for contribution of participants in an SIA process to contribute to the redesign of the project or plan or in decision-making about what constitutes appropriate development (Vanclay 2002a, 2003a).

The name debate

There has been debate over the precise meanings of terms such as social impact assessment, social impact analysis, social analysis, social assessment, social appraisal, socio-economic impact assessment, social monitoring, social soundness analysis, social soundness assessment and social risk assessment. The term 'human impact assessment' is also emerging as an area that integrates social with health impact assessment. Much of this debate has been ill-informed, failing to appreciate the different levels of meaning of SIA, or not realising that definitions change over time. New terminology has emerged: in response to changing ideologies; to promote a particular ideological position; for the requirements of specific institutions such as the World Bank; or to escape previous poor practice. But a rose is a rose by any other name. Inventing new terms doesn't resolve the situation and, worse, it inevitably creates confusion.

Within the International Association for Impact Assessment, the analysis of social issues is called 'Social Impact Assessment'. SIA, as a term, has been formally in existence and used in textbooks, newsletters and conferences since the early 1970s. However, other terms are used to describe largely similar activities. Some (e.g. Taylor et al. 1995) use the term 'Social Assessment' because of their desire to overcome the negative connotation implied by the word 'impact'. They also see the process as being more widely applicable than just within a regulatory process, but they see it as fundamentally similar to the description of SIA given in this chapter and in the *International Principles for Social Impact Assessment*. Lane and colleagues (2001: 3), however, consider 'social assessment' as being broader than SIA. They see social assessment as being 'a means for collecting and organising information about the social domain in a way that informs natural resource decision-making', in other words much more than analysing, monitoring and managing the social consequences of planned interventions.

Various institutions have their own proclivities. The US Department of Transport, for example, uses the term 'community impact assessment' to mean SIA, albeit in its traditional understanding. Within the World Bank, 'Social Assessment' developed initially as a way of including social considerations into the design of projects, largely by participatory processes. 'Social Analysis' developed later as an analytical process in the consideration of social issues in the design of a project and as a mechanism for considering the broader social context in which development occurs. They are now developing 'social impact analysis'. Other multilateral development agencies (and banks) tend to emulate the World Bank and use the same terms, but not always with the same meanings. The United States Agency for International Development (USAID) – which has the dubious mission statement of 'furthering America's foreign policy interests in expanding democracy and free markets' (USAID 2004: website) – tends to use the term 'Social Soundness Analysis'. For them, Social Soundness Analysis has three dimensions:

> (1) the compatibility of the activity with the sociocultural environment in which it is to be introduced (its sociocultural-feasibility); (2) the likelihood that the new practices or institutions introduced among the initial activity target population will be diffused among other groups (i.e., the spread effect); and (3) the social impact or distribution of benefits and burdens among different groups, both within the initial activity population and beyond. (USAID n.d.: 1)

The term 'socio-economic impact assessment' is popular with various governments and with some corporations, but is generally rejected by SIA professionals because where it is used, social considerations tend to become subordinate to economic ones. 'Social monitoring' tends to be favoured in the mining industry, again presumably as a way of keeping the process firmly in the control of the industry and of limiting the role that the social contribution might have.

There has been widespread ignorance of the field of impact assessment in general, and of SIA in particular. This has been partly responsible for the growth in terminology, especially with the institutionalisation of people working within various agencies oblivious to the professional developments happening in the world around them. The multitude of terms is not helpful to the discipline as a whole, nor is the fractured understanding of the issues. The community of scholars will only benefit when they rally around a common unifying entity, that is, social impact assessment, and when they share a commitment to a common understanding of that entity, such as that reflected in the *International Principles for Social Impact Assessment*. In this respect, the failure of the Interorganizational Committee (2003) to embrace the new understanding of SIA in its 2003 revision of the USA Guidelines and Principles is disappointing.

The what is 'social' debate

A major issue in impact assessment generally relates to what issues are included under the rubric of 'social impacts'. Consultants who undertake impact assessments are limited to what is specified in their Terms of Reference, but they also need to be mindful of what constitutes acceptable professional practice, duty of care considerations, and to some extent professional culture. How 'social impacts' are interpreted is therefore central to what is actually considered.

Unfortunately, the case history of impact assessment presents a poor record, reflecting inadequate consideration of social issues. There are many reasons for this, including the asocietal mentality (an attitude that humans don't count) that exists, and the lack of SIA expertise (see Burdge & Vanclay 1995; Vanclay 1999; Lockie 2001). Too often the only impacts that are considered are economic impacts and demographic changes. A further problem is that there are groups with narrow sectoral interests who have advocated new fields of impact assessment. A limited view of what is 'social' creates demarcation problems about what impacts are to be identified by SIA, versus what is considered by fields such as health impact assessment, cultural impact assessment, heritage impact assessment, aesthetic impact assessment, or gender impact assessment. The SIA community of practitioners considers that all issues affecting people, directly or indirectly, are pertinent to SIA.

SIA is thus best understood as an umbrella or overarching framework that embodies the evaluation of all impacts on humans and on the ways in which people and communities interact with their sociocultural, economic and biophysical surroundings. SIA thus encompasses a wide range of specialist sub-fields involved in the assessment of areas such as aesthetic impacts (landscape analysis), archaeological and cultural heritage impacts (both tangible and non-tangible), community impacts, cultural impacts, demographic impacts, development impacts, economic and fiscal impacts, gender impacts, health and mental health impacts, impacts on indigenous rights, infrastructural impacts, institutional impacts, leisure and tourism impacts, political impacts (human rights, governance, democratisation etc.), poverty, psychological and psychosocial impacts, resource issues (access and ownership of resources), impacts on social and human capital, and other impacts on societies. As such, comprehensive SIA cannot be undertaken by a single person; it requires a team approach.

Elsewhere, I have identified over eighty social impact concepts that should be considered in SIA (Vanclay 2002b). A convenient way of conceptualising social impacts is as changes to one or more of the following (Vanclay 2002a: 389):

- *people's way of life* – that is, how they live, work, play and interact with one another on a day-to-day basis;
- *their culture* – that is, their shared beliefs, customs, values and language or dialect;
- *their community* – its cohesion, stability, character, services and facilities;
- *their political systems* – the extent to which people are able to participate in decisions that affect their lives, the level of democratisation that is taking place, and the resources provided for this purpose;
- *their environment* – the quality of the air and water people use; the availability and quality of the food they eat; the level of hazard or risk, dust and noise they are exposed to; the adequacy of sanitation, their physical safety, and their access to and control over resources;
- *their health and wellbeing* – where 'health' is understood in a manner similar to the World Health Organization definition: 'a state of complete physical, mental, and social [and spiritual] wellbeing and not merely the absence of disease or infirmity';
- *their personal and property rights* – particularly whether people are economically affected, or experience personal disadvantage which may include a violation of their civil liberties;
- *their fears and aspirations* – their perceptions about their safety, their fears about the future of their community, and their aspirations for their future and the future of their children.

While there is an emerging field of gender impact assessment (see Srinivasan & Mehta 2003), the view in the SIA field is that the gendered nature of impacts is something that is well known in SIA and does not need to be considered as a separate process. Having too many separate processes only sets one form of impact assessment against another, reduces the funding that is available for each, and leads to marginalisation and to confusion in the results, outcomes and solutions. It also risks annoying the community.

The policy-level debate

Although EIA and SIA were always intended to be applicable at the policy level as well as the project level (Taylor et al. 1995; Becker 1997), their practice and experience have been at the project level. EIA and SIA have failed to consider the positive outcomes of development, have not considered the goals of development, and in the case of SIA, has tended to emphasise the impacts on individuals rather than on societies as a whole. Cumulative effects have been ill-considered, and upstream and downstream impacts and the second and higher-order impacts ignored. Boothroyd (1995: 90) identifies four reasons why EIA and SIA have been constrained to the project level:

- Projects are tangible, dramatic, highly organized, discrete geographically and temporally . . .
- Localized negative impacts of individual projects are perceived to be able to be mitigated, or are regarded as insignificant in comparison to project benefits, thus EIA can be applied rigorously on a case-by-case basis without threatening the current unsustainable development path and those most benefiting from it.
- Policy-making is secretive, or at least guarded. Power-holders feel threatened by increasing explicit systematization and the public accountability it produces or seems to imply . . .
- The most important policies are unwritten, often unspoken, certainly not reviewed, and are therefore not assessable.

To counter the deficiencies in EIA, Strategic Environmental Assessment (SEA) has emerged as 'a form of impact assessment that can assist managers and leaders in policy, planning and programmatic decisions' (Partidário 2000: 647). Its definition is much discussed, with many writers shying away from a precise definition (see Partidário 1999; Verheem & Tonk 2000).

Although SEA is an important advance, doubt remains about the extent to which it provides anything different to project-level thinking, and the extent to which it involves social as well as biophysical impacts. SEA was spawned from the technical EIA paradigm by people ignorant of the established field of policy assessment (Boothroyd 1995) and who largely ignored the SIA field.

While the benefits of moving upstream to the policy level are recognised, they have focused on the assessment of consequences, rather than the design of policy (Boothroyd 1995). SIA at the policy level must provide a basis for considering what policies are applicable to solve a particular problem, and not just what problems arise from a specific policy, as Boothroyd (1998: web document) indicates:

> Unfortunately, as a supplement to policy evaluation and analysis, impact assessment is still deficient in four respects. Like policy analysis, impact assessment tends to be reductionist – it focuses on discrete, measurable impacts (e.g. loss of traditional foods by people dislocated by industrial development) rather than on whole dynamic systems (which could include inter-related traditional foods and traditional cultures and sources of meaning). Impact assessment is conducted primarily on the most concrete policy issues, that is, project approvals (e.g. mines) and sometimes programs (mining exploration), avoiding the general policies (frontier expansion) which drive projects and programs. Impact assessment as practised is almost exclusively concerned with the externalities of efficiency-oriented physical projects: the externalities of a trade regulation, or of a new literacy-oriented school curriculum, for example, are rarely assessed.
>
> To the extent impact assessment of macro-policies is conducted at all, it is as an impotent add-on to policy analysis, just as physical project impact assessment has always been an add-on to engineering and financial cost–benefit analysis. As a side-show to the policy analysis main event, impact assessment has little influence on decisions. The main question still is, 'will the policy achieve the initiating objective and implicit larger societal goal?' – not, 'what are the consequences of this policy for society?'

The assessment of policy is conceptually different from assessment at the project level. SIA is unlikely to influence policy even when there is genuine desire to include social considerations and significant resources are allocated (Fisher 2001).

FUTURE DIRECTIONS

The SIA field has a sure future. There will always be interest in the social consequences of planned interventions, and in the ways of improving planned interventions to minimise unplanned harm and maximise good. But the field of SIA will continue to have a troubled time. Policy-making is a political process, not a technical one. The contribution of SIA will often be at odds with political interests. The prevailing asocietal mentality and the anti-democratic forces in society will actively seek to reduce the contribution of SIA. Various governments in Australia and elsewhere have established SIA Units only to close them down later when they realised that the advice was not what they wanted to hear (see Dale et al. 2001).

The SIA field will nevertheless persist and strengthen. With codification through the International Association for Impact Assessment, there will be less fluctuation due to ideological swings, although there is always room for the field to mature, to gain practical experience, and to build theory. Fashions come and go, and current fashions, such as the risk society, social capital and the triple bottom line, will have to be contested or accommodated if SIA is not to be subsumed or made redundant or irrelevant (Vanclay 2003b). With continued community dialogue, there could be an ongoing demand for SIA to be part of the policy process. With continued calls for democratisation, the asocietal mentality will be reduced, and greater realisation of the potential contribution of SIA will occur.

Discussion Questions

1. What are the three levels of SIA?
2. What is the definition of the new conceptualisation of SIA, and what is significant about this definition in terms of the origin of SIA?
3. How does SIA differ from EIA?
4. What is the asocietal mentality and how does it affect SIA?
5. What are the differences in meaning between the terms Public Consultation, Public Participation and Public Involvement, and why are these differences important?
6. What is a 'stakeholder', and what problems are associated with the use of this term?
7. At what phases in the lifecycle of a planned intervention can SIA be used?
8. What are some of the difficulties in applying SIA at the policy level?
9. Consider the different meanings of 'the environment' and the politics associated with these.
10. Outline the SIA critique of the field of strategic environmental assessment.

Glossary of Terms

Affected publics: the sum total of all individuals, communities and groups of people who will be affected by a planned intervention, whether directly or indirectly.

Cumulative impacts: There are three dimensions relating to cumulative impacts. One relates to the aggregation of small impacts. The second relates to synergistic and/or catalytic interaction between impacts. The third relates to flow-on effects, or higher-order effects, rather than the immediate impacts.

First-order impact (or direct impact): an impact that occurs as a direct result of the planned intervention. In SIA, it refers to social changes and social impacts caused directly by the intervention itself. First-order impacts also cause second and higher-order or indirect impacts, those that occur as a result of another change caused by a planned intervention.

Interested and affected parties (or publics):   a concept with a meaning similar to 'stakeholders' but which avoids the popular misunderstanding associated with the latter term: that a stakeholder has to have a vested interest in the planned intervention.

Mitigation: the process of devising and implementing processes, procedures and changes to a planned intervention or additional activities in order to reduce or eliminate the negative impacts likely to be experienced.

Public consultation, public involvement, and public participation:

> 'Public involvement' is the overarching concept: public involvement is a process for involving the public in the decision-making process . . . Such involvement can be brought about through a spectrum of activities ranging from consultation to participation, the key difference being the degree to which those involved in the process are able to influence, share or control the decision-making. While 'Consultation' includes education, information sharing, and negotiation – the goal being better decision making by the organization consulting the public – 'Participation' actually brings the public into the decision-making process. Typically, public involvement has focused primarily on consulting the public with no options for greater participation. Arnstein (1969) was one of the first to identify the 'ladder of citizen participation', which ranged from persuasion at the one end of the spectrum to self-determination at the other end. This is the strongest form of public participation where the process is directly undertaken by the public with the proponent accepting the outcome. (Roberts 2003: 259–60)

Social Impact Assessment: 'includes the processes of analysing, monitoring and managing the intended and unintended social consequences, both positive and negative, of planned interventions (policies, programs, plans, projects) and any social change processes invoked by those interventions. Its primary purpose is to bring about a more sustainable and equitable biophysical and human environment' (IAIA 2003: 2).

Stakeholders: all those individuals and groups who have a vested interest in, are affected by (whether negatively or positively) or have the power to affect or influence a planned intervention.

References

Arnstein, S. R. 1969, 'A ladder of citizen participation', *American Institute of Planning Journal* 35(4): 216–24.
Australia 1999, *Environment Protection and Biodiversity Conservation Act 1999* (No. 91, 1999).
Australia 2000a, *Gene Technology Act 2000* (No. 169, 2000).

Australia 2000b, *Gene Technology Bill 2000 Explanatory Memorandum*.

Becker, H. 1997, *Social Impact Assessment: Method and Experience in Europe, North America and the Developing World*, London: University College London Press.

Becker, H., and F. Vanclay (eds) 2003, *The International Handbook of Social Impact Assessment*, Cheltenham, UK: Edward Elgar.

Boothroyd, P. 1995, 'Policy assessment'. In Vanclay and Bronstein, *Environmental and Social Impact Assessment*, pp. 83–126.

Boothroyd, P. 1998, *Social Policy Assessment Research: The establishment, the underground*, Ottawa: International Development Research Centre Working Series Paper #5. http://www.idrc.ca/socdev/pub/documents/ spar.html (accessed 20 January 2004).

Burdge, R.J., and F. Vanclay 1995, 'Social Impact Assessment'. In Vanclay and Bronstein, *Environmental and Social Impact Assessment*, pp. 31–66.

COAG (Council of Australian Governments) 1997, *Principles and Guidelines for National Standard Setting and Regulatory Action by Ministerial Councils and Standard-Setting Bodies*, Canberra: Department of Prime Minister and Cabinet. http://www.dpmc.gov.au/pdfs/coagpg.pdf (accessed 20 January 2004).

Dale, A., C.N. Taylor and M. Lane (eds) 2001, *Social Assessment in Natural Resource Management Institutions*, Melbourne: CSIRO.

Egre, D., and P. Senecal 2003, 'Social impact assessments of large dams throughout the world', *Impact Assessment and Project Appraisal* 21(3): 215–24.

Fisher, M. 2001, 'Hastening the evolution of Australian rural industry policy'. In Dale et al., *Social Assessment in Natural Resource Management Institutions*, pp. 231–41.

Harvey, N. 1998, *Environmental Impact Assessment: Procedures, practice and prospects in Australia*, Melbourne: Oxford University Press.

IAIA (International Association for Impact Assessment) 2003, *International Principles for Social Impact Assessment*, Fargo ND: IAIA. www.iaia.org/Publications/SP2.pdf (accessed 20 January 2004). Also published as F. Vanclay 2003, 'International Principles for Social Impact Assessment', *Impact Assessment and Project Appraisal* 21(1): 5–11.

Interorganizational Committee on Guidelines and Principles for Social Impact Assessment 1994, 'Guidelines and principles for Social Impact Assessment', *Impact Assessment* 12(2): 107–52.

Interorganizational Committee on Principles and Guidelines for Social Impact Assessment 2003, 'Principles and Guidelines for Social Impact Assessment in the USA', *Impact Assessment and Project Appraisal* 21(3): 231–50.

Lane, M., A. Dale and N. Taylor 2001, 'Social assessment in natural resource management'. In Dale et al. *Social Assessment in Natural Resource Management Institutions*, pp. 3–12.

Lockie, S. 2001, 'SIA in Review', *Impact Assessment and Project Appraisal* 19(4): 277–88.

Partidário, M. 1999, 'Strategic environmental assessment: principles and potential'. In J. Petts (ed.) *Handbook of Environmental Impact Assessment*, London: Blackwell, pp. 12–32.

Partidário, M. 2000, 'Elements of an SEA framework', *Environmental Impact Assessment Review* 20(6): 647–63.

Queensland 1990, *Local Government (Planning and Environment) Act 1990.*

Roberts, R. 2003, 'Involving the public'. In Becker and Vanclay, *International Handbook*, pp. 258–77.

Slootweg, R., F. Vanclay and M. van Schooten 2001, 'Function evaluation as a framework for the integration of social and environmental impact assessment', *Impact Assessment and Project Appraisal* 19(1): 19–28.

Srinivasan, B., and L. Mehta 2003, 'Assessing gender impacts'. In Becker and Vanclay, *International Handbook*, pp. 161–78.

Taylor, C.N., C.H. Bryan and C.G. Goodrich 1995, *Social Assessment: Theory, process and techniques*, 2nd edn, Christchurch, NZ: Taylor Baines & Associates.

USAID (United States Agency for International Development) n.d., *Social Soundness Analysis* (ADS Supplementary Reference 202). http://www.usaid.gov/policy/ads/200/2026s7.pdf (accessed 20 January 2004).

USAID 2004, *About USAID.* http://www.usaid.gov/about_usaid/index.html (accessed 20 January 2004).

Vanclay, F. 1999, 'Social Impact Assessment'. In J. Petts (ed.) *Handbook of Environmental Impact Assessment*, vol. 1, Oxford: Blackwell Science, pp. 301–26.

Vanclay, F. 2002a, 'Social Impact Assessment'. In M. Tolba (ed.) *Responding to Global Environmental Change*, Chichester, UK: Wiley, pp. 387–93.

Vanclay, F. 2002b, 'Conceptualising social impacts', *Environmental Impact Assessment Review* 22(3): 183–211.

Vanclay, F. 2003a, 'Conceptual and methodological advances in Social Impact Assessment'. In Becker and Vanclay, *International Handbook*, pp. 1–10.

Vanclay, F. 2003b, 'Experiences from the field of Social Impact Assessment: where do TBL, EIA and SIA fit in relation to each other?'. In B. Pritchard, A. Curtis, J. Spriggs and R. Le Heron (eds) *Social Dimensions of the Triple Bottom Line in Rural Australia*, Canberra: Bureau of Rural Sciences, pp. 61–80.

Vanclay, F., and D. Bronstein (eds) 1995, *Environmental and Social Impact Assessment*, Chichester, UK: Wiley

Verheem, R., and J. Tonk 2000, 'Strategic environmental assessment: one concept, multiple forms', *Impact Assessment and Project Appraisal* 18(3): 177–82.

CHAPTER SIXTEEN

CRIMINOLOGY, SOCIAL REGULATION AND ENVIRONMENTAL HARM

Rob White

The interface of criminology, as a discrete field of study, with environmental issues in a manner that involves concerted professional attention and hands-on intervention has been forcefully advocated by Lynch & Stretsky in a recent article.

> In general, criminologists have often left the study of environmental harm, environmental laws and environmental regulations to researchers in other disciplines. This has allowed little room for critical examination of individuals or entities who/which kills, injures and assaults other life forms (human, animal or plant) by poisoning the earth. In this light, a green criminology is needed to awaken criminologists to the types of major environmental harm and damage that can result from environmental harms; the conflicts that arise from attempts at defining environmental crime and deviance; and the controversies still raging over possible solutions, given extensive environmental regulations already in place (Lynch & Stretsky 2003: 231).

Sociologically speaking, any attempt to take up the challenge offered here will require rigorous and sophisticated analysis of the social dynamics that shape and allow certain types of activities harmful to the environment (and human beings and animals) to take place over time. This sort of analysis, in turn, demands that environmental issues be framed within the context of a sociological, and one might add criminological, imagination (see Wright Mills 1959; White 2003). That is, study must appreciate the importance of situating environmental harm as intrinsically socially and historically located and created. Interpretation and analysis thus has to be mindful of how current trends reflect the structure of global/local societies, the overall direction in which such societies are heading, and the ways in which diverse groups of people are being affected by particular social, economic and political processes.

BACKGROUND TO THE ISSUES

There are major long-standing issues relating to how 'harm' (and indeed 'crime') is to be defined in criminological terms, and of what the responses to harm should consist. The usual divide is usually seen in terms of those who adopt a strict legal-procedural approach to defining harm, and those who opt for a broader socio-legal approach. The former is basically dependent on legal definitions that proscribe certain actions in law (see Tappan 1947). The latter allows for investigation of phenomena (such as white-collar crime and denial of human rights) by adoption of conceptions of harm which are not limited to definitions solely generated by the state (see Sutherland 1949; Green & Ward 2000). The conundrum of definition is made worse in the specific area of environmental harm in that many of the most serious forms of such harm in fact constitute 'normal social practice' and are quite legal even if environmentally disastrous.

The politics of definition are further complicated by the politics of 'denial', in which particular concrete manifestations of social injury and environmental damage are obfuscated, ignored or redefined in ways which re-present them as being of little relevance to either academic criminological study or state criminal justice intervention. In a manner analogous to the denial of human rights violations (see Cohen 2001), environmental issues call forth a range of techniques of neutralisation on the part of nation-states and corporations that ultimately legitimate and justify certain types of environmentally unfriendly activities. For example, this takes the form of 'greenwashing' media campaigns that misconstrue the nature of collective corporate business in regard to the environment (Beder 1997). It involves attacking and delegitimating the arguments of critics of particular kinds of biotechnological development (see Hannigan 1995; Hindmarsh 1996; Hager & Burton 1999). For governments, denial of harm is usually associated with economic objectives and the appeal to forms of 'sustainable development' that fundamentally involve further environmental degradation (see Hessing 2002; White 2002).

The emergence of environmental or green criminology in recent years has been marked by efforts to reconceptualise the nature of 'harm' in a more inclusive manner (see Chunn et al. 2002; White 2003). Much of this work has been directed at exposing different instances of substantive environmental injustice and ecological injustice. It has also involved critique of the actions of nation-states and transnational capital for fostering particular types of harm, and for failing to adequately address or regulate harmful activity. Drawing on a wide range of ideas and empirical materials, recent work dealing with environmental harm has ventured across a range of concerns.

- It has documented the existence of law-breaking with respect to pollution, disposal of toxic waste and misuse of environmental resources (Pearce & Tombs 1998).
- It has raised questions relating to the destruction of specific environments and resources, in ways which are 'legal' but ecologically very harmful to plants, animals and human beings (Halsey 1997a).
- It has emphasised the dynamic links between social inequality and the distribution of environmental 'risk', particularly as these affect poor and minority populations (Bullard 1994; Low & Gleeson 1998).
- It has investigated the specific place of animals in relation to issues of 'rights' and human–non-human relationships on a shared planet (Benton 1998).
- It has criticised the inadequacies of environmental regulation in both philosophical and practical terms (Gunningham et al. 1995; Harvey 1996; Halsey & White 1998).
- It has reconsidered the nature of victimisation, including social and governmental responses to this victimisation (Williams 1996).

Given the pressing nature of many environmental issues, it is not surprising that many criminologists are now seeing environmental crime and environmental victimisation as areas for concerted analytical and practical attention – as areas of work that require much more conceptual development and empirical attention (see South 1998; Lynch & Stretsky 2003).

KEY DEBATES

Generally speaking, a specifically criminological concern with environmental harm can be seen to centre around four major questions. The first two are: what is the nature, extent and impact of environmental harm; and who are the main victims of environmental harm? These questions go to the heart of the debates over defining 'crime' and 'harm' within an environmental framework. The second set of questions include: what responses are or should be taken to address environmental harm, particularly from the point of view of social regulation; and how best should the relationship between human beings and the environment be managed? These questions address issues of social control and the exercise of state and class power in preventing or responding to actual and potential instances of environmental harm.

Defining environmental harm

The first question is basically one of definition (and thus intrinsically, explanation). Initially it can be said that to define what constitutes environmental harm implies a particular philosophical stance on the relationship between human beings and nature. What is 'wrong' or 'right' environmental practice very much depends on the criteria used to conceptualise the values and interests represented in this relationship, as reflected for instance in anthropocentric, biocentric, and ecocentric perspectives (see Halsey & White 1998). The notion of harm is thus inherently ideological. It is constructed in practice through ongoing political contestation over the meaning and interpretation of specific acts and omissions.

This is also reflected in discipline-specific debate over how and under what conditions an act or omission might be conceived as an environmental 'crime'. A strict legalist approach tends to focus on the central place of criminal law in the definition of criminality. Thus, as Situ & Emmons (2000: 3) see it: 'An environmental crime is an unauthorised act or omission that violates the law and is therefore subject to criminal prosecution and criminal sanctions.' Here we see a strictly legal interpretation of 'harm', one that intrinsically relies on and reflects state definitions of environmental crime, regardless of ecological processes and outcomes.

Other writers, however, argue that, as with criminology in general, the concept of 'harm' ought to encapsulate those activities that may be legal and 'legitimate' but that nevertheless negatively impact on people and environments (Lynch & Stretsky 2003). This broader conceptualisation of crime or harm is deemed to be essential in evaluating the systemic, as well as particularistic, nature of environmental harm. To put it differently, it is important to distinguish (and make the connection between) specific instances of harm arising from imperfect operation (such as pollution spills), and systemic harm which is created by normatively sanctioned forms of activity (clearfelling of Amazon forests). The first is deemed to be 'criminal' or 'harmful', and thus subject to social control. The second is not. The overall consequence of this is for the global environmental problem to get worse, in the very midst of the proliferation of a greater range of regulatory mechanisms, agencies and laws. This is partly ingrained in the way in which environmental risk is compartmentalised: specific events or incidents attract sanction, while wider legislative frameworks may set parameters on, but nevertheless still allow, other ecologically harmful practices to continue.

The state nevertheless has a formal role and commitment to protect citizens from the worst excesses or worst instances of environmental victimisation. Hence the introduction of extensive legislation and regulatory procedures designed to give the appearance of active intervention, and the implication that laws exist which actually do deter such harms. The existence of such laws may be encouraging in that they reflect historical and ongoing struggles over certain types of activity. But how or whether they are used once again begs the questions of the relationship between the state and the corporate sector, and the capacity of business to defend its interests by legal and extra-legal means. Much of the critical environmental criminology attempts to expose the class-based, gendered and racialised origins and effects of environmental harms, and thereby to illustrate the circuits of power that lie behind the harms that occur (Chunn et al. 2002).

Part of the practical problem in defining environmental harm revolves around knowledge of the harm. Part of the problem also lies in the complexity of causation when it comes to certain types of harm. Many businesses, for example, can gain protection from close public or state surveillance through the very processes of commercial negotiation and transaction. These range from appeals to 'commercial confidentiality' through to constraints associated with the technical nature of evidence required. For example, there is often difficulty in law in assigning 'cause' in cases of environmental harm due to the diffuse nature of responsibility for particular effects, such as pollution in an area of multiple producers (e.g. mining companies). Furthermore, it has been pointed out that 'evidence frequently can only be collected through the use of powers of entry, the ability to take, analyse and interpret appropriate samples and a good knowledge of the processes or activities giving rise to the offence' (Robinson 1995: 13). Such powers impinge on the 'private' property rights and commercial interests which are at the heart of the capitalist political economy.

Current regulatory apparatus, informed by the ideology of 'sustainable development', is largely directed at bringing ecological sustainability to the present mode of producing and consuming – one based on the logic of growth, expanded consumption of resources, and the commodification of more and more aspects of nature (see White 2002). At the heart of these processes is a culture that takes for granted, but rarely sees as problematic, the proposition that continued expansion of material consumption is both possible and will not harm the biosphere in any fundamental way. Some aspects of denial are consciously and directly linked to instrumental purposes (as in firm or industry campaigns to de-legitimate environmental action surrounding events or developments that are manifestly harmful to local environments). At a more general level, however, denial is ingrained in the hegemonic dominance of anthropocentric, and specifically capitalist, conceptions of the relationship between human beings and nature. Basic assumptions about economic growth and commodity production – central components of the dominant worldview – make it difficult for many people to see the essence of the problem as lying in the system itself (see Harvey 1996; Halsey & White 1998).

Determining environmental victims

Environmental victimisation can be defined as specific forms of harm which are caused by acts (e.g. dumping of toxic waste) or omissions (e.g. failure to provide safe drinking water) leading to the presence or absence of environmental agents (e.g. poisons, nutrients) which are associated with human injury (see Williams 1996). The management of these forms of victimisation is generally retrospective (after the fact), and involves a variety of legal and social responses.

From the point of view of environmental criminology, analysis of the nature of environmental harm has to take into account objective and subjective dimensions of victimisation. It also has to locate the processes of environmental victimisation within the context of the wider political economy. That is, the dynamics of environmental harm cannot be understood apart from consideration of who has the power to make decisions, the kinds of decisions that are made, in whose interests they are made, and how social practices based on these decisions are materially organised. Issues of power and control have to also be analysed in the light of global economic, social and political developments.

Analysis of environmental issues proceeds on the basis that someone or something is indeed being harmed. In this regard, a distinction is sometimes made between 'environmental justice' and 'ecological justice'. *Environmental justice* refers to the distribution of environments among peoples (Low & Gleeson 1998) and the impacts of particular social practices on specific populations. The focus of analysis therefore is on human health and well-being and how these are affected by particular types of production and consumption. Here we can distinguish between environmental issues that affect everyone, and those that disproportionately affect specific individuals and groups (see Williams 1996). Water, for example, is a basic human requirement. The people who have been affected by poor water quality in the advanced capitalist countries represent a broad cross-section of the population (see White 2003). In this sense, the periodic water crises are non-discriminatory in terms of class, gender, ethnicity and so on. There is thus a basic 'equality of victims', in that some environmental problems threaten everyone in the same way, as in the case for example of ozone depletion, global warming, air pollution and acid rain (Beck 1996).

Nevertheless, as extensive work on specific incidents and patterns of victimisation demonstrates, some people are more likely to be disadvantaged by environmental problems than others. For instance, American studies have identified disparities involving many different types of environmental hazards that adversely affect people of colour throughout the United States (Bullard 1994). Other work in Canada and Australia has focused on the struggles of indigenous people to either prevent the environmental degradation of their lands or to institute their own methods of environmental protection (see Langton 1998; Rush 2002). It is clear that environmental racism manifests itself in a number of different guises and is directly related to social inequalities that are, in turn, reflected in particular forms of environmental victimisation (Stephens 1996). The specificity of those placed at greater or disproportionate risk from environmental harm is also reflected in literature that acknowledges the importance of class, occupation, gender, and more recently, age, in the construction of special environmental interest groups (Stephens 1996; Williams 1996; Chunn et al. 2002). There are thus patterns of 'differential victimisation' that are evident with respect to the siting of toxic waste dumps, extreme air pollution, access to safe, clean drinking water and so on.

Another dimension of differential victimisation relates to the subjective disposition and consciousness of the people involved. The specific groups who experience environmental problems may not always describe or see the issues in strictly environmental terms.

> In our communities, the smell coming from sewage plants was never perceived as an environmental issue but as a survival issue ... In workplaces, when workers are being poisoned or contaminated ... we do not refer to them as environmental issues but as labour issues. Again, the same thing for farmworkers and the issue of pesticides. In the '60s and '70s, there was organizing around the lead-based paints used in housing projects. When the paint curled up and chipped off, children in the projects were eating it and getting sick. When we dealt with this issue, we perceived it as an issue of tenants' rights. (Moore 1990: 16)

The unequal distribution of exposure to environmental risks, whether it be in relation to the location of toxic waste sites or proximity to clean drinking water, may not always be conceived as an 'environmental' issue, or indeed as an environmental 'problem'. For instance, Harvey (1996) points out that the intersection of poverty, racism and desperation may occasionally lead to situations where, for the sake of jobs and economic development, community leaders actively solicit the relocation of hazardous industries or waste sites to their neighbourhoods.

Ecological justice refers to the relationship of human beings generally to the rest of the natural world. The focus of analysis therefore is on the health of the biosphere, and more specifically plants and creatures that also inhabit the biosphere (see Benton 1998; Franklin 1999). The main concern is with the quality of the planetary environment (which is frequently seen to possess its own intrinsic value) and the rights of other species (particularly animals) to life free from torture, abuse and destruction of habitat. Specific practices, and choices, in how humans interact with particular environments present immediate and potential risks to everything within them. For example, the practice of clearfelling old-growth forests directly affects many animal species by destroying their homes (see Halsey 1997b). Similarly, local natural environments, and non-human inhabitants of both wilderness and built environments, are negatively affected by human practices that destroy, rechannel or pollute existing fresh-water systems.

The 'choices' ingrained in environmental victimisation (of human beings, of the non-human world) stem from systemic imperatives to exploit the planetary environment for production of commodities for human use. This is not a politically neutral process. In other words, how human beings produce, consume, and reproduce themselves is socially patterned in ways that are dominated by global corporate interests (see Athanasiou 1996; White 2002). The dominance of neo-liberal ideology as a guiding rationale for further commodification of nature, and the concentration of decision-making in state bureaucracies and transnational corporate hands, are seen to accelerate the rate and extent of environmental victimisation (Hessing 2002). The power of capitalist hegemony manifests itself in the way in which certain forms of production and consumption become part of a taken-for-granted commonsense, as the experiences and habits of everyday life (White 2002). Environmental and ecological injustices are rationalised as the way things have to be in the new globalised system of trade and governance. Dissent may be tolerated, but only in so far as it does not challenge the core imperatives for economic growth, or the generalised power of transnational capital to set the agenda.

Responding to environmental harm

There are three main approaches to the analysis and study of environmental regulation. One is to chart existing environmental legislation
and to provide a sustained socio-legal analysis of specific breaches of law,
the role of law enforcement agencies, and the difficulties and opportunities of using criminal law against environmental offenders (see del Frate
& Norberry 1993; Gunningham et al. 1995; Heine et al. 1997; Situ &
Emmons 2000). For those who view environmental harm in a wider
lens than that provided by criminal law, this approach has clear limitations. In particular, the focus on criminal law, regardless of whether or
not the analysis is critical or confirming, offers a rather narrow view of
'harm' that can obscure the ways in which the state facilitates destructive environmental practices and environmental victimisation.

Accordingly, much academic work has shifted towards a second
approach, one in which the key focus is not criminal sanctions as
such, but regulatory strategies that might be used to improve environmental performance. Here the main concern is with varying forms
of 'responsive regulation' (Ayres & Braithwaite 1992; Braithwaite
1993) and 'smart regulation' (Gunningham & Grabosky 1998). These
approaches attempt to recast the state's role by using non-government,
and especially private sector, participation and resources in fostering
regulatory compliance in relation to the goal of 'sustainable development'. Increasingly important to these discussions is the perceived and
potential role of third-party interests, in particular non-government
environmental organisations, in influencing policy and practice
(Gunningham & Grabosky 1998; Braithwaite & Drahos 2000; O'Brien
et al. 2000). The main concern of this kind of approach is with reform
of existing methods of environmental protection. The overall agenda
of writers in this genre has been summarised as follows: 'Generally
speaking, environmental reformers are optimistic about the possibilities of addressing environmental harms without fundamentally changing the status quo. Either implicitly or tacitly, minimization ("risk
management") rather than elimination of environmental depredation
is conceived as the reformist object' (Chunn et al. 2002: 12).

In the third approach, writers tend to be more sceptical of the previous perspectives and developments, arguing that many key elements of
such strategies dovetail with neo-liberal ideologies and practices (especially the trend towards deregulation of corporate activity) in ways
that will not address systemic environmental degradation (White 1999;
Snider 2000). Furthermore, from the point of view of the restructuring of class relationships on a global scale, reforms in environmental

management and regulation are seen to be intrinsically linked to the efforts of transnational corporations to further their hegemonic control over the planet's natural resources (see Goldman 1998a, b; Pearce & Tombs 1998; White 2002). In this type of analysis, political struggle and the contest over class power are viewed as central to any discussion of environmental issues. Issues of gender, ethnicity, and race are important to these discussions as well, but are incorporated into a specifically eco-socialist understanding of capitalism and nature (see Pepper 1993; O'Connor 1994; Chunn et al. 2002).

Rather than focusing on the notions of effectiveness, efficiency, and the idea of win–win regulatory strategies, this approach is concerned with social transformation (Chunn et al. 2002). As such, the analysis proceeds from the view that critical analysis must be counter-hegemonic to dominant hierarchical power relationships, and that present institutional arrangements require sustained critique and systemic change. It has been pointed out, for instance, that the broad tendency under neo-liberalism has been towards deregulation (or, as a variation of this, 'self-regulation') when it comes to corporate harm and wrongdoing. In the specific area of environmental regulation, the role of government remains central, even if only by the absence of state intervention. The general trend has been away from direct governmental regulation and towards 'softer' regulatory approaches.

Snider (2000), for example, describes how in Canada, despite policy directives specifying 'strict compliance', a permissive philosophy of 'compliance promotion' has reigned. Given the tone of mainstream regulation literature (which offers a theoretical justification for enlisting private interests through incentives and inducements), it is hardly surprising that persuasion is favoured at the practical level. But, more than seeing this as simply a reflection of the new regulatory ideology, it is essential to consider the financial and political environment within which regulators are forced to work. For example, while never before in history have there been so many laws pertaining to the environment, it is rare indeed to find extensive government money, resources and personnel being put into enforcement and compliance activities. Rather, these are usually provided in the service of large corporations, as a form of state welfare designed to facilitate and enhance the business climate and specific corporate interests.

Self-regulation and environmental management

The role of criminologists in providing a theoretical cover for questionable environmental practices is an issue warranting serious consideration, particularly in relation to contemporary thinking about corporate regulation. In general, the idea of encouraging trustworthiness ('virtue') by individual companies and by industry associations – of promoting regulation by 'consent' – has, unsurprisingly, gained a modicum of support within official government circles and among business leaders. The mainstream (and dominant) model of regulation is based on the notion of a regulatory pyramid, with persuasion the favoured approach at the base moving upwards to coercion at the pinnacle (Ayres & Braithwaite 1992; Grabosky 1994, 1995). The basic argument has been that the most effective regulatory regime is one that combines a range of measures, in most of which the targeted institutions and groups are meant to have some interest in participating, or complying with (see Braithwaite 1993; Gunningham & Grabosky 1998). The implication is that corporate attitudes should be the focus for reform, including cases where third-party input is encouraged in the regulatory arena.

In the specific area of environmental regulation, there is likewise support for the idea that persuasion, not coercion, is or ought to be the key regulatory mechanism. This is usually associated with the ideology of 'self-regulation' (see e.g. Grabosky 1994, 1995). Here it is argued that corporate regulation should be informed by the idea of enlisting 'private interests' in regulatory activity via 'inducements' such as adopting waste minimisation programs which translate into more efficient production, or earning a good reputation among consumers for environmental responsibility. Such perspectives also reinforce the notion that 'markets' are and should be a key component of any regulatory system. This appears to tap into the dominant ideological framework of capitalism – that 'the market knows best'. Analytically, the problem of regulation is divorced from structural analysis of political economic relations. Rather, great emphasis is placed on 'illustrative' examples and case studies in which specific forms of incentive and compliance appear to be 'working' in an environmentally friendly manner (see Grabosky 1994). Much is thus made of how 'market opportunities' can drive 'environmentally appropriate commercial activity'. Less is said about the overall expansionary pressures of capitalism, or the immediate pressures on particular businesses to curb environmental controls precisely because of competitive costs (Haines 1997; White 2002).

At the centre of changes to environmental regulation has been the movement towards 'corporate ownership' of the definitions of, and responses to, environmental problems. This has taken different forms. One type of response, for example, has been to adopt the language of 'environmental management systems' (EMS) and to assert that regulation is best provided by those industries and companies directly involved in production processes. This occurs both at the level of particular firms and in relation to the setting of international standards for environmental management, as in the case of the ISO 14000. Those criminologists who broadly support 'smart regulation' strategies that combine a variety of regulatory measures also tend to be supportive of EMS types of regulatory systems. There are various dimensions to EMS, relating to environmental valuation and risk analysis, product design, corporate culture and environmental awareness, supply and waste chain management, and so on (see Kirkland & Thompson 1999). While EMS may be seen as progressive and a positive step forward in environmental regulation by some (especially proponents of 'smart regulation' strategies), embedded within EMS ideology are certain assumptions that imply 'more of the same' rather than system transformation.

This is acknowledged in literature that is more sceptical and critical of what EMS appears to offer. For example, Levy (1997) observes that EMS does address some of the worst environmental excesses (i.e. real material consequences of production practices in specific cases). But, he argues, at an ideological and symbolic level EMS primarily serves to construct products and companies as 'green' and to legitimise corporate management as the primary societal agent responsible for addressing environmental issues. That is, decisions to adopt EMS can be seen as part of a political, practical and ideological response to the threat to corporate hegemony posed by environmental movements. According to the critics, the key message of EMS is that corporations have the know-how to protect the environment best (on our behalf), in that they have the technical means and managerial strategies to do so. As Levy (1997) points out, and as echoed in the 'smart regulation' literature (see Gunningham & Grabosky 1998), EMS is presented as a win–win opportunity in which the potential structural conflicts between profit maximisation and environmental goals are avoided.

As well as not being demonstrated empirically, this provides yet another cover to circumvent government regulation. Much the same has been argued in relation to the 'standards' put forward by the International Standards Organisation (ISO). That is, the ISO 14000 (relating to environmental impacts) constitutes a private sector initiative that allows for the state to divest itself of regulatory functions and simultaneously remove regulation and standards-setting from the democratic process and beyond the reach of citizens and social movements (Wall & Beardwood 2001). The issue of who regulates what, and who controls the process, is central to any discussion of how best to respond to environmental harm.

For those interested in social transformation, the need is to deconstruct notions such as 'self-regulation' by examining the real world of corporate activity, and the persisting damage caused by systemic exploitation of human beings and the natural environment. This involves identifying and explaining the transformations in regulation along a number of dimensions, taking into account the specific role of international capitalist institutions such as the World Trade Organisation, the International Monetary Fund and the World Bank, and accounting for the shifts in regulatory emphasis away from the state and towards private business interests (see e.g. Goldman 1998a, b; White 1999; O'Brien et al. 2000).

In the context of neo-liberal policies and globalised capital relations, the relationship of the state to private interests is ultimately contingent on baseline economic criteria. Where environmental harm has occurred, for example, there are a number of issues which impinge on the capacity and willingness of the state to enforce compliance or prosecute wrongdoing. Some of these are threats of litigation by companies against the state or third-party critics on the basis of 'commercial reputation'; a paucity of independent scientific expertise (related to cuts in the number of state regulators, the buying off of experts by companies, and funding crises affecting the research direction of academic institutitions); the complexities associated with investigation and action in relation to transnational corporate environments (e.g. formation of international cartels, potential threats to future investment, monopolisation of particular industries such as water); and state reluctance to enforce compliance due to ideological attachments to privatisation and corporatisation, and the notion of less state intervention the better (see White 1998).

FUTURE DIRECTIONS

By its very nature, the development of environmental criminology as a field of sustained research and scholarship will incorporate many different perspectives and strategic emphases. For some, the point of academic concern and practical application will be to reform aspects of the present system. Critical analysis, in this context, will consist of thinking of ways to improve existing methods of environmental regulation and perhaps to seek better ways to define and legally entrench the notion of environmental crime. For others, the issues raised above are inextricably linked to the project of social transformation. From this perspective, analysis ought to focus on the strategic location and activities of transnational capital, as supported by hegemonic nation-states on a world scale, and it ought to deal with systemic hierarchical inequalities. Such analysis is seen to open the door to identifying the strategic sites for resistance, contestation and struggle on the part of those fighting for environmental and ecological justice.

There are many variations in how people will approach environmental issues. For example, specific orientations include 'soft green' approaches supportive of sustainable development through to 'hard green' approaches calling for ecological sustainability. In the end, it is clear that there are major political divisions within the broad spectrum of academic work (and indeed of green movements), and these have major implications for whether action will be taken in collaboration with capitalist institutions, or whether it will be directed towards radically challenging these institutions. This too – the process of reform and transformation (incorporating questions of legislative change, and legal and illegal forms of activism) – is an area requiring further critical investigation by environmental criminologists.

Discussion Questions

1. What are the major concerns of a 'green criminology'?
2. Should environmental harm be defined solely in terms of the criminal law?
3. What kinds of issues does the recent environmental criminology concern itself with?
4. How does environmental harm or degradation impact on different social groups?
5. In what sense might there be a basic 'equality of victims' when it comes to certain types of environmental harm?
6. Provide an example of a state-sanctioned practice that does harm to the environment.
7. What are the main differences between a 'smart regulation' approach and an approach that is concerned with social transformation?
8. Why is 'persuasion' limited when it comes to environmental regulation of corporate activity?
9. Who is the main accountable party in cases where EMS has been adopted?
10. What do you think are the key issues that arise from sociological analysis of the regulation of environmental harm?

Glossary of Terms

Environmental Management System: a form of 'self-regulation' in which individual firms or industry groups take on the role of regulating how specific processes are to be organised, with the intention of designing environmentally friendly systems of production.

Ecological justice refers to the relationship of human beings generally to the rest of the natural world, and includes concerns relating to the health of the biosphere, and more specifically plants and creatures that also inhabit the biosphere.

Environmental harm: a wide variety of injuries and degradations linked to the use, misuse and poor management of the 'natural environment', including such things as pollution, toxic waste and killing of plants, soils, and animals.

Environmental justice: the distribution of environments among peoples in terms of access to and use of specific natural resources in defined geographical areas, and the impacts of particular social practices and environmental hazards on specific populations (e.g. as defined on the basis of class, occupation, gender, age, ethnicity).

Environmental victimisation: the social processes in which specific forms of harm are caused by 'acts' such as dumping of toxic waste, or 'omissions' such as failure to provide safe drinking water, leading to the presence or absence of environmental agents (e.g. poisons, nutrients) which are associated with human injury.

Green criminology: Also known as 'environmental criminology', green criminology refers to the study of environmental harm, environmental laws and environmental regulation by criminologists.

Smart regulation: a particular strand of environmental regulatory theory which argues that the best way to regulate environmental activity is to bring together a wide diversity of methods and actors in order to foster compliance in relation to the goal of 'sustainable development'.

Social regulation: the processes by which a society, via the state or private agencies, monitors and shapes how human activity is to be carried out in relation to specific political, economic and social agendas (which are, in turn, reflective of particular social interests).

References

Athanasiou, T. 1996, *Divided Planet: The ecology of rich and poor*, Boston MA: Little, Brown & Co.

Ayres, I., and J. Braithwaite 1992, *Responsive Regulation: Transcending the deregulation debate*, New York: Oxford University Press.

Beck, U. 1996, 'World risk society as cosmopolitan society? Ecological questions in a framework of manufactured uncertainties', *Theory, Culture, Society* 13(4): 1–32.

Beder, S. 1997, *Global Spin: The corporate assault on environmentalism.* Melbourne: Scribe Publications.

Benton, T. 1998, 'Rights and justice on a shared planet: more rights or new relations?', *Theoretical Criminology* 2(2): 149–75.

Boyd, S., D. Chunn and R. Menzies (eds) 2002, *Toxic Criminology: Environment, law and the state in Canada*, Halifax, Canada: Fernwood Publishing.

Braithwaite, J. 1993, 'Responsive business regulatory institutions'. In C. Coady and C. Sampford (eds) *Business Ethics and the Law*, Sydney: Federation Press.

Braithwaite, J., and P. Drahos 2000, *Global Business Regulation*, Cambridge University Press.

Bullard, R. 1994, *Unequal Protection: Environmental justice and communities of color*, San Francisco CA: Sierra Club Books.

Chunn, D., S. Boyd and R. Menzies 2002, ' "We all live in Bhopal": criminology discovers environmental crime'. In Boyd et al., *Toxic Criminology*.

Cohen, S. 2001, *States of Denial: Knowing about atrocities and suffering*, Oxford: Polity/Blackwell.

del Frate, A., and J. Norberry (eds) 1993, *Environmental Crime: Sanctioning strategies and sustainable development*, Rome: UNICRI /Sydney: Australian Institute of Criminology.

Franklin, A. 1999, *Animals and Modern Cultures: A sociology of human–animal relations in modernity*, London: Sage.

Goldman, M. 1998a, 'Introduction: The Political Resurgence of the Commons'. In M. Goldman (ed.) *Privatizing Nature: Political struggles for the global commons*, London: Pluto Press in association with Transnational Institute.

Goldman, M. 1998b, 'Inventing the commons: theories and practices of the commons' professional'. In Goldman, *Privatizing Nature*.

Grabosky, P. 1994, 'Green markets: environmental regulation by the private sector', *Law and Policy* 16(4): 419–48.

Grabosky, P. 1995, 'Regulation by reward: on the use of incentives as regulatory instruments', *Law and Policy* 17(3): 256–79.

Green, P., and T. Ward 2000, 'State crime, human rights, and the limits of criminology', *Social Justice* 27(1): 101–15.

Gunningham, N., and P. Grabosky 1998, *Smart Regulation: Designing environmental policy*, Oxford: Clarendon Press.

Gunningham, N., J. Norberry and S. McKillop (eds) 1995, *Environmental Crime: Conference proceedings*, Canberra: Australian Institute of Criminology.

Hager, N., and B. Burton 1999, *Secrets and Lies: The anatomy of an anti-environmental PR campaign*, New Zealand: Craig Potton Publishing.

Haines, F. 1997, *Corporate Regulation: Beyond 'punish or persuade'*, Oxford: Clarendon Press.

Haines, F. 2000, 'Towards understanding globalisation and control of corporate harm: a preliminary criminological analysis', *Current Issues in Criminal Justice* 12(2): 166–80.

Halsey, M. 1997a, 'Environmental crime: towards an eco-human rights approach', *Current Issues in Criminal Justice* 8(3): 217–42.

Halsey, M. 1997b, 'The wood for the paper: old-growth forest, hemp and environmental harm', *Australian and New Zealand Journal of Criminology* 30(2): 121–48.

Halsey, M., and R. White 1998, 'Crime, ecophilosophy and environmental harm', *Theoretical Criminology* 2(3): 345–71.

Hannigan, J. 1995, *Environmental Sociology: A social constructionist perspective*, London: Routledge.

Harvey, D. 1996, *Justice, Nature and the Geography of Difference*, Oxford: Blackwell.

Heine, G., M. Prabhu and A. del Frate (eds) 1997, *Environmental Protection: Potentials and limits of criminal justice*, Rome: UNICRI.

Hessing, M. 2002, 'Economic globalization and Canadian environmental restructuring: the mill(ennium)-end sale'. In Boyd et al., *Toxic Criminology*.

Hindmarsh, R. 1996, 'Bio-policy translation in the Public Terrain'. In G. Lawrence, K. Lyons and S. Momtaz (eds) *Social Change in Rural Australia*, Rockhampton, Qld: Rural Social and Economic Research Centre, CQU.

Kirkland, L.-H., and D. Thompson 1999, 'Challenges in designing, implementing and operating an environmental management system', *Business Strategy and the Environment* 8: 128–43.

Langton, M. 1998, *Burning Questions: Emerging environmental issues for indigenous peoples in northern Australia*, Darwin: Centre for Indigenous Natural and Cultural Resource Management.

Levy, D. 1997, 'Environmental management as political sustainability', *Organization and Environment* 10(2): 126–47.

Low, N., and B. Gleeson 1998, *Justice, Society and Nature: An exploration of political ecology*, London: Routledge.

Lynch, M., and P. Stretsky 2003, 'The meaning of green: contrasting criminological perspectives', *Theoretical Criminology* 7(2): 217–38.

Moore, R. 1990, 'Environmental Inequities', *Crossroads* No. 1.

O'Brien, R., A. Goetz, J. Scholte and M. Williams 2000, *Contesting Global Governance: Multilateral economic institutions and global social movements*, Cambridge University Press.

O'Connor, J. 1994, 'Is sustainable capitalism possible?'. In M. O'Connor (ed.) *Is Capitalism Sustainable? Political economy and the politics of ecology*, New York: Guilford Press.

Pearce, F., and S. Tombs 1998, *Toxic Capitalism: Corporate crime and the chemical industry*, Aldershot, UK: Dartmouth Publishing Co.

Pepper, D. 1993, *Eco-Socialism: From deep ecology to social justice*, New York: Routledge.

Robinson, B. 1995, 'The nature of environmental crime'. In Gunningham et al., *Environmental Crime*.

Rush, S. 2002, 'Aboriginal resistance to the abuse of their natural resources: the struggle for trees and water'. In Boyd et al., *Toxic Criminology*.

Situ, Y., and D. Emmons 2000, *Environmental Crime: The criminal justice system's role in protecting the environment*, Thousand Oaks CA: Sage.

Snider, L. 2000, 'The sociology of corporate crime: an obituary (or: whose knowledge claims have legs?)', *Theoretical Criminology* 4(2): 169–206.

South, N. 1998, 'A green field for criminology', *Theoretical Criminology* 2(2): 211–34.

Stephens, S. 1996, 'Reflections on environmental justice: children as victims and actors', *Social Justice* 23(4): 62–86.

Sutherland, E. 1949, *White Collar Crime*, New York: Dryden Press.

Tappan, P. 1947, 'Who is the criminal?' *American Sociological Review* 12: 96–102.

Wall, E., and B. Beardwood 2001, 'Standardizing globally, responding locally: the new infrastructure, ISO 14000, and Canadian agriculture', *Studies in Political Economy* 64: 33–58.

White, R. 1998, 'Environmental criminology and Sydney water', *Current Issues in Criminal Justice* 10(2): 214–19.

White, R. 1999, 'Criminality, risk and environmental harm', *Griffith Law Review* 8(2): 235–57.

White, R. 2002, 'Environmental harm and the political economy of consumption', *Social Justice* 29(1–2): 82–102.

White, R. 2003, 'Environmental issues and the criminological imagination', *Theoretical Criminology* 7(4): 483–506.

Williams, C. 1996, 'An environmental victimology', *Social Justice* 23(4): 16–40.

Wright Mills, C. 1959, *The Sociological Imagination*, New York: Oxford University Press.

INDEX

For EU product safety concerns, contact us at Calle de José Abascal, 56–1°,
28003 Madrid, Spain or eugpsr@cambridge.org.

www.ingramcontent.com/pod-product-compliance
Ingram Content Group UK Ltd.
Pitfield, Milton Keynes, MK11 3LW, UK
UKHW042211180425
457623UK00011B/168